PROBING THE ATOM

PROBING THE ATOM

INTERACTIONS OF COUPLED STATES,
FAST BEAMS, AND LOOSE ELECTRONS

Mark P. Silverman

PRINCETON UNIVERSITY PRESS   PRINCETON, NEW JERSEY

Library of of Congress Cataloging-in-Publication Data

Silverman, Mark P.
Probing the atom : interactions of coupled states, fast beams,
and loose electrons / Mark P. Silverman.
p. cm.
Includes bibliographical references and index.
ISBN 0-691-00962-7 (cloth : alk. paper)
1. Atomic structure. I. Title.
QC173.4.A87 S55 2000
539′.14—dc21 99-31032 CIP

This book has been composed in Times Roman

The paper used in this publication meets the minimum requirements
of ANSI/NISO Z39.48-1992 (R 1997) (*Permanence of Paper*)

http://pup.princeton.edu

Printed in the United States of America

1 3 5 7 9 10 8 6 4 2

—————————To Susan, Chris, and Jen—————————

AND IN MEMORY OF

Frank Pipkin

MENTOR, COLLABORATOR, FRIEND

Contents

"IF, IN SOME cataclysm, all of scientific knowledge were to be destroyed, and only one sentence passed on to the next generations of creatures, what statement would contain the most information in the fewest words?" So began Richard Feynman's three-volume grand tour of physics that has excited, delighted, and edified many a physicist, including me. "I believe," Feynman answered, "it is the *atomic hypothesis*...that *all things are made of atoms—little particles that move around in perpetual motion, attracting each other when they are a little distance apart, but repelling upon being squeezed into one another*. In that one sentence...there is an enormous amount of information about the world, if just a little imagination and thinking are applied."[1] (Feynman's italics)

Feynman's query and reply attest to the far-reaching significance of the discreteness of matter. But how do we know that atoms exist—that is, as real entities and not just as a hypothetical conjecture? I have frequently asked physics classes that question, only to be met with vacant stares and shrugged shoulders. It is, in fact, not a trivial question—and even reputable chemists and physicists (like Wilhelm Ostwald and Ernst Mach) were disbelievers until well past the turn of the twentieth century. The answer—before tunneling and other forms of microscopy provided images of individual atoms—lay in the multitude of experimental ways to determine Avogadro's number: for example, combination of Faraday's constant with electron charge, volume of helium production from radioactive decay of radium, Rayleigh scattering of light, mean free path in gases, and Brownian motion displacement in liquids, to cite but a few. There is no question that matter is discrete.

Recognition of the existence of atoms, however, is by no means the end of an inquiry into the atomic structure of matter, but only the barest of beginnings. Of what are atoms made? What holds the pieces together? How do these constituents move? Where are they located? These and many more such questions fixed the direction of physics research during the first few decades of the twentieth century. Though the frontiers of physics have since shifted to other domains, the internal workings of the atom and the interactions of atoms with one another and with various kinds of electromagnetic radiation are by no means exhausted subjects.

In the search for answers, the use of rf (radiofrequency and microwave) oscillating fields and atomic beams has provided—and continues to provide—a penetrating glimpse into the finest details of atomic structure and behavior. The physical principles and practical implementation of this profoundly important mode of interrogating nature are the subject of this book. Although a self-contained and independent work, *Probing the Atom* is, in a sense, the concluding sequel to two earlier books expounding facets of my scientific research and educational interests. In *More than One Mystery: Explorations of Quantum Interference*,[2] I discussed theoretical and experimental studies related to the fundamentals of quantum mechanics. In *Waves and Grains: Reflections on Light and Learning*,[3] I wrote of my investigations in physical optics. Although optics and quantum physics are certainly essential parts of atomic spectroscopy, the subject matter discussed in the present volume is an entirely separate component of my scientific work with respect to both experimental techniques and theoretical inquiries.

I began my investigations in atomic physics in the 1960s at a time when precision measurements of atomic structure—especially that of hydrogen and helium—provided some of the most demanding tests of quantum electrodynamics (QED), the most successful physical theory then known. Subsequently unified with the weak interactions, the "electroweak" theory is still unrivaled in its predictive capacity. For me personally, this was in a number of ways a time of marked change and new beginnings. Physically, I had traveled over three thousand miles to take up research and teaching at Harvard. Professionally, it was also at this juncture that I finally crossed over from chemistry to physics, the subject that actually interested me the most since childhood. And last, as a card-carrying physicist (APS membership card), I had the good fortune to be "in at the beginnings"—to borrow the apt title of Philip Morse's autobiography[4]—with regard to the combined use of particle accelerators and rf fields for probing the atom.

In the evolution of atomic physics, the investigations of Willis Lamb and his students stand out in my mind as of epochal significance. Thermally dissociating molecular hydrogen to obtain a continuous beam of individual atoms, and driving transitions between close-lying atomic states by means of an external oscillating field, Lamb and Retherford conclusively demonstrated in 1947 the inadequacy of Dirac's theory to account for a small, but critical, feature of the energy level structure of hydrogen: the nondegeneracy of $2^2S_{1/2}$ and $2^2P_{1/2}$ states. The energy interval, known as the "Lamb shift," between fine structure states[5] of the same J quantum number is purely quantum electrodynamical in origin. This discovery, and a suite of experiments performed with other students, stimulated intense theoretical efforts to wring from QED ever more precise predictions. These experiments became the historic prototypes from which followed numerous variants for measuring the fine and hyperfine structure of atoms.

As a young scientist searching for an original and productive line of research, I was deeply influenced by Lamb's ingenious experiments. Although it was from Feynman's papers that I came to understand quantum electrodynamics, it was largely Lamb's papers that taught me atomic physics. Painstakingly photocopied and collected in a black three-ring binder, these papers accompanied me everywhere, and I read them in every spare moment.

During my years as a chemist investigating complex molecules, I had become familiar with a variety of resonance techniques—nuclear magnetic resonance, electron spin resonance, electron-nuclear double resonance—as well as diverse kinds of optical spectroscopy. (Quite possibly I may have been the first to try electron-nuclear-optical *triple* resonance, but my initial attempt was not successful and, for various reasons, the experiment had to be abandoned.) At the same time, I had strong theoretical interests, especially in the electromagnetic interactions of atoms and molecules, which I explored from a perspective of quantized fields similar to the approach being developed in France at the time under the colorful designation *l'atome habillé* ("the dressed atom"). With this experimental and theoretical background I formally became an atomic physicist and, together with F. M. Pipkin and C. W. Fabjan, undertook the introduction of fast beams into the rf spectroscopy of atoms.[6]

The advantage of an atomic beam is that atoms are conveyed from the region where they are created (and in which perturbing fields and particles are ordinarily present) to a separate, tranquil environment for spectroscopic probing. Unless atoms are transported rather quickly, however, excited states will decay away, and only ground or long-lived metastable states will remain to be examined. The atoms of a thermal beam move relatively slowly with wide dispersion in atomic velocity. At 2500 K, the temperature at which H_2 molecules were dissociated in Lamb's experiments, the most probable speed of an atom was about 8×10^5 cm/s. A hydrogen atom excited into the $3D$ state therefore would have to survive on average for over 400 times the state mean lifetime (15.6 ns) in order to cover a distance of 5 cm. The probability (e^{-400}) of finding such states at 5 cm from their source is very low.

With a beam of atoms moving at least 250 times faster, however, a large number of short-lived states become available for study. This is approximately the orbital speed of an electron in the hydrogen ground state. Such translational speeds can be achieved with an ion accelerator by passing fast ions—protons in the case of a hydrogen beam—through an electron-rich material such as a thin solid foil or extended gas target where they capture electrons to become rapidly moving neutral atoms. This is the basis of what I have called the SABER—*S*pectroscopy with *A*ccelerated *B*eam and *E*lectric *R*esonance—method.

Besides an accelerator, the SABER method incorporated another feature relating to electric resonance spectroscopy that marked a radical departure from earlier practices. The term "electric resonance" refers to the increasing effectiveness of an oscillating field to induce electric-dipole transitions between two quantum states in the measure that the applied frequency of the oscillator matches the Bohr frequency of the states. Ideally, a resonance lineshape is "swept out" by varying the applied frequency and monitoring an experimental signal related to the transition probability. From location of the resonance, usually determined at the peak or, in some cases, the valley of a lineshape, the energy level separation of coupled states of interest can be determined.

Although conceptually the most obvious way to proceed, this ideal method had not been implemented before because the oscillator power output and transmission line characteristics changed with frequency. Even available power monitors were frequency sensitive. Had Lamb conducted his experiments by varying the rf oscillator frequency, the resulting lineshapes would have been distorted and of little practical value.

To circumvent this problem, Lamb kept the radiofrequency of his oscillator fixed and swept, instead, the strength of an external static magnetic field which, by means of the Zeeman effect, smoothly varied the energy level separation of the coupled states. But the price of immersing the atomic beam in a static magnetic field was a high one, for it necessitated (in Lamb's words) "a lengthy programme of calculations and measurements" to correct for a large number of systematic effects that distorted or shifted the lineshape. Among these were variation of the transition matrix element with magnetic field, the production of motional electric fields with resulting lineshifts (Stark effect) and differential quenching of hyperfine states, unsymmetrical distribution of the hyperfine levels about the mean fine structure energy due to incomplete decoupling of the nuclear angular momentum (Back-Goudsmit effect), and curvature of the Zeeman energy levels due to partial decoupling of the orbital and spin angular momenta (Paschen-Back effect).

In the SABER method, use of a rf power meter to monitor oscillator power accurately, and carefully designed radiofrequency and microwave chambers, made it practicable to generate electric resonance lineshapes by the ideal procedure of varying the oscillator frequency in the absence of external static fields. (The magnetic field of the Earth itself was nulled by surrounding the apparatus with three mutually orthogonal sets of current loops, called Helmholtz coils.) In this way resonance lineshapes of atoms in short-lived states could be measured under well-understood and controlled conditions.

Generally speaking, there are traditionally two principal objectives to the study of atoms, as pursued by two different categories of experiments: One objective is to understand the internal dynamics of an atom from its energy level structure; this is the province of spectroscopy. The other is to study the dynamics of atoms interacting with one another (or with basic constituents like the electron, proton, or alpha particle); this ordinarily falls within the domain of collision physics. There is, as well, a third objective—not necessarily mutually exclusive of the first two, but distinct enough to warrant separate mention—in which an atom is a convenient system to demonstrate fundamental consequences of quantum physics. One might be interested, for example, in observable manifestations of coherence in the superposition of state amplitudes or the seemingly odd nonlocal properties of widely separated, correlated particles.

Here the focus of attention lies not so much with the particular atom under study, but rather in the opposition of the laws of the microscale world to the familiar expectations of Newtonian and Maxwellian physics.

This book is concerned essentially with the first two objectives (as the third is the focus of attention in *More than One Mystery*). Like its two predecessors, it is neither a textbook nor a compendious survey of a subfield of physics. Rather, by concentrating on the well-defined topic, drawn from my own researches, of atoms in transient states interacting with oscillating fields and giving rise to spontaneously emitted light, this book addresses broad conceptual and practical matters concerning the investigation of atomic structure and behavior.

But the book was not written only for spectroscopists. Indeed, anyone with an interest in applications of quantum physics, in mathematical methods of solving quantum problems, and in design strategies for testing the predictions of quantum mechanics in the laboratory will, I hope, also find this book useful reading.

For example, to extract information from almost any spectroscopic method of probing atoms and molecules one must solve a fundamental theoretical problem: the interaction of a quantum system of two or more states with an oscillating field. Surprisingly (perhaps), the Schrödinger equation for this system is not amenable to exact closed-form analytical solution even in the apparently simplest case of a two-level atom and monochromatic linearly polarized field. In this book I examine the atom-field interaction at different levels of complexity and refinement, each level illuminating the physical processes involved from a different perspective and yielding complementary results.

This progression of treatments begins with the rotating-wave approximation (RWA) to the two-level atom devised by I. I. Rabi (for the magnetic resonance of stable nuclear spin states) and then presents a much better approximation, applicable to transient states, which I have called the oscillating-field theory (OFT).[7] The RWA provides an exact solution to a simplified problem in which the oscillating field is replaced by a field rotating about the quantization axis. By contrast, the OFT solves the two-state Schrödinger equation for a true oscillating field by a suite of transformations that yields results in close agreement with exact numerical integration.

Next I develop a "generalized resonant field" (GRF) theory for the driven multilevel (≥ 2) atom[8] and apply this theory to two-, three-, and four-level systems (which are the couplings that arise in atomic hydrogen in the absence of external static fields). Although a generalization of the rotating-wave approximation for a two-state system, the theory in the case of electric resonance does not really involve rotating (i.e., circularly polarized) fields; rather, the procedure constitutes a form of gauge or phase transformation.

In the preceding two stages of development, the external driving field was treated classically. In the third stage I quantize the rf field (as well as the atom), consider the coupling of states by multiple-quantum transitions, and examine the correspondence between the two analytical approaches.[9] This dual quantization leads to a graphically expressive interpretation of the atom-field interaction and approximations permitting a significant reduction in complexity of the resulting equations of motion for transitions of a given photon number.

In the last stage of refinement, the optical field (rather than the rf field) is quantized, and the radiative decay process, which had hitherto been treated phenomenologically, is examined from the perspective of quantum electrodynamics.[10] The results go beyond the pioneering studies of natural linewidth by Weisskopf and Wigner, applicable in the absence of external fields, and show how a rf field can significantly affect both the temporal and spectral properties of the spontaneous emission lineshape. Such variations underlie the optical detection of rf-induced transitions in so-called "bottle experiments," whereby both the stimulation and detection of transitions occur within the confines of a resonance cell.

Besides a comprehensive examination of the atom–rf field interaction, I consider as well the general principles behind a number of other issues essential to creating and probing atoms, such as the production, acceleration, focusing, and charge-exchange conversion of ions, the design and testing of the rf system, use of rf fields to select atomic states and to narrow resonance lineshapes, and the detection and interpretation of the optical signals.[11]

One of the novel aspects of atomic spectroscopy discussed in this book is the use of rf lineshapes to obain information relating, not only to energy level structure, but also to the interactions by which neutral atoms are created by electron capture. This information is contained in the entire shape of the resonance line, as deduced in the first five chapters of the book. Using the formalism of the density matrix (or statistical operator) developed in chapter 4, I illustrate by means of a simple model (Coulombic spin-independent collision model) how to account for a panorama of resonance lineshapes by means of a small number of collision parameters from which follow the relative populations of all substates within an electronic manifold.[12]

The preparation of *Probing the Atom* was greatly stimulated by invited lectures on various aspects of quantum theory and atomic physics that I gave throughout the 1990s. I am especially thankful to colleagues and students at the Arizona State University, University of Canterbury (New Zealand), University of Connecticut, Helsinki University of Technology, University of Helsinki, University of Maine, and University of Missouri for their kind reception and thought-provoking questions.

I also very much appreciate the efforts of Dr. Trevor Lipscombe, Physics Editor at Princeton University Press, who enthusiastically supported this project, Dr. Jennifer Slater who copyedited the manuscript, my son Chris Silverman who designed the front cover, and my wife Dr. Susan Brachwitz who designed the periodic table in figure 1.1.

Finally, one comment about the title of this book, which, as the foregoing discussion amply justifies, is all about coupled states, fast beams, and loose (i.e., capturable) electrons. While at Harvard many years ago, I attended an informal seminar by Jim (*The Double Helix*) Watson who, in the course of his lecture, remarked that he initially intended to call his book *Base Pairs*, but chose, instead, the less suggestive title by which it subsequently gained fame (or notoriety). I am not unaware that the subtitle of this book lends itself to certain word play—but so be it; physicists have often boldly gone where biologists feared to tread.

NOTES

1. R. P. Feynman, R. B. Leighton, and M. Sands, *The Feynman Lectures on Physics* (Addison-Wesley, Reading, MA, 1963), pp. 1–2.

2. Mark P. Silverman, *More than One Mystery: Explorations in Quantum Interference* (Springer, New York, 1995).

3. Mark P. Silverman, *Waves and Grains: Reflections on Light and Learning* (Princeton University Press, Princeton, NJ, 1998).

4. Philip M. Morse, *In at the Beginnings: A Physicist's Life* (MIT Press, Cambridge, 1977).

5. In the standard spectroscopic notation $n^{2S+1}L_J$ of atomic terms, n is the principal quantum number designating the electronic manifold, and S, L, J are respectively the spin, orbital, and total angular momentum quantum numbers. For hydrogen, $S = 1/2$ leading to a "multiplicity" $2S + 1 = 2$.

6. C. W. Fabjan, F. M. Pipkin, and M. P. Silverman, "Radiofrequency Spectroscopy of Hydrogen Fine Structure in $n = 3$, 4, 5," *Phys. Rev. Lett.* **26** (1971):347.

7. M. P. Silverman and F. M. Pipkin, "Interaction of a Decaying Atom with a Linearly Polarized Oscillating Field," *J. Phys. B: Atom. Molec. Phys.* **5** (1972):704.

8. M. P. Silverman and F. M. Pipkin, "Optical Electric Resonance Investigation of a Fast Hydrogen Beam, Part I: Theory of the Atom RF Field Interaction," *J. Phys. B: Atom. Molec. Phys.* **7** (1974):704.

9. M. P. Silverman, *Optical Electric Resonance Investigation of a Fast Hydrogen Beam* (Ph.D. Thesis, Harvard University, 1973).

10. M. P. Silverman and F. M. Pipkin, "Radiation Damping of Atomic States in the Presence of an External Time Dependent Potential," *J. Phys. B: Atom. Molec. Phys.* **5** (1972):2236.

11. M. P. Silverman and F. M. Pipkin, "Optical Electric Resonance Investigation of a Fast Hydrogen Beam, Part II: Theory of the Optical Detection Process," *J. Phys. B: Atom. Molec. Phys.* **7** (1974):730.

12. M. P. Silverman and F. M. Pipkin, "Optical Electric Resonance Investigation of a Fast Hydrogen Beam, Part III: Experimental Procedure and Analysis of H($n = 4$) Quantum States," *J. Phys. B: Atom. Molec. Phys.* **7** (1974):747.

PROBING THE ATOM

CHAPTER 1

Energies and Spectral Lines

1.1 ANATOMY OF HYDROGEN

In the beginning (if there was a beginning), there was hydrogen...and to a lesser extent helium. Perhaps not at the *very* beginning (when God only knows what there was), but close enough for the purposes of an atomic physicist. Not only is hydrogen the most abundant element in the Universe, it is the simplest naturally occurring bound state system,[1] and consequently one of few that permit detailed and precise comparison with theoretical models. The spectrum of hydrogen has long served as a touchstone for all theories of atomic structure. Likewise, in the investigation of collisional processes, hydrogen is the preeminent model system for a wide variety of atomic interactions critical to atomic physics, atmospheric physics, astrophysics, and nuclear fusion, to cite but a few. No wonder the physicist's view of the periodic table may have resembled figure 1.1 throughout much of the twentieth century.

With but one orbiting electron (of charge $-e$ and mass m), hydrogen is paradoxically both the archetype of all atoms as well as the least typical atom. As an archetype, the structure of hydrogen defines the scale of atomic size (Bohr radius $a_0 = \hbar^2/me^2 \sim 0.053$ nm) and atomic energy (Rydberg $R_\infty = e^2/2a_0 \sim 2.2 \times 10^{-11}$ erg), atomic units of electrical potential ($e/a_0 \sim 27.2$ V) and field strength ($e/a_0^2 \sim 5.1 \times 10^9$ V/cm), the electron orbital speed relative to that of light (Sommerfeld fine structure constant $\alpha = e^2/\hbar c \sim 1/137$), and other dynamical quantities. (Recall that $\hbar \equiv h/2\pi$, where h is Planck's constant.) The set of hydrogen "spin orbitals" or wavefunctions is ordinarily the starting point for the *aufbau* ("buildup") of the electron shells of the elements and the chemical bonds of molecules. On the other hand, hydrogen is the only element for which there are no interactions amongst electrons nor need to invoke the Pauli exclusion principle. Highly excited alkali atoms like sodium, for example, may have a gross energy level structure that resembles hydrogen, but upon closer scrutiny—no matter how high

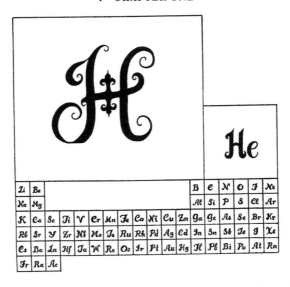

Figure 1.1 The atomic physicist's view of the periodic table.

the degree of excitation of the single valence electron—the pattern of energy levels always departs from hydrogen, sometimes in spectacular ways.[2] The reason is that these atoms still have an inner electronic core within which electrons are perturbed by one another and by the nucleus, whereas hydrogen does not. Since the "three-body" problem (classical or quantum) is in general not amenable to closed-form solution, there is no other atom, not even helium, for which an analytical model can be solved exactly even in the absence of electron spin and the effects of special relativity and quantum electrodynamics.

The apparent simplicity of the hydrogen atom, however, tends to diminish with continued scrutiny. Indeed, a point has been reached where the bound electron energy can be predicted with greater precision only if the internal structure of the proton—a small dynamical world of its own with internal forces scarcely reaching to one hundred thousandth of a Bohr radius—is better known.

The energy spectrum of an atom uniquely reflects the interactions of its constituent particles. It is therefore somewhat like a mirror in which the image is scrambled. To decode that image and see clearly how it came to be—in effect, to follow the path from hypothetical force laws to observable spectrum—is one of the essential goals of atomic physics. To good approximation, the energy E_0 of a hydrogen atom (in the absence of external fields) can be expressed as the sum of four contributions:

$$E_0(nLJF) = E_D(nJ) + E_{hf}(nLJF) + E_N + E_{QED}, \qquad (1)$$

each of which will be examined in sequence. The quantum numbers labeling the state are the principal quantum number n designating the electronic manifold, electron orbital angular momentum L, total electron (orbital + spin) angular momentum J, and total atomic (electron + nuclear) angular momentum F. In the absence of external fields, the atom is in a spherically symmetric environment, and the energy of a state is independent of orientation relative to the (arbitrary) axis of quantization as specified by a magnetic quantum number M_F.

The first term on the right hand side, which takes the compact form[3]

$$E_D = \frac{mc^2}{\sqrt{1 + \left(\dfrac{\alpha}{n - \delta}\right)^2}} \qquad (2a)$$

with

$$\delta = J + \tfrac{1}{2} - \sqrt{(J + \tfrac{1}{2})^2 - \alpha^2}, \qquad (2b)$$

is the Dirac energy for an electron attracted to an immobile point nucleus of charge $+1$ (in units of e). [For a nucleus with atomic number Z, one replaces α by $Z\alpha$ in Eqs. (2a, b).] Since α/n is interpretable as the orbital speed v/c of an electron in the nth manifold, Eq. (2a) almost resembles the classical relativistic expression for kinetic energy (corrected by a small "quantum defect" δ)—except that the sign within the radical is wrong; the expression multiplying mc^2 is not the relativistic factor $\gamma = 1/\sqrt{1 - (v/c)^2}$. The Dirac term actually encompasses *all* the energy contributions of a charged point mass with spin $\tfrac{1}{2}\hbar$: the electron kinetic and coulombic potential energies (including the rest-mass energy mc^2 which is the same for all eigenstates) as well as the spin-orbit interaction leading to the level fine structure.

To see the individual terms explicitly that arise from the simpler nonrelativistic Schrödinger equation with spin added (that is, the Pauli equation), one can expand Eqs. (2a, b) in a power series in α to obtain the familiar expression

$$E_D(nJ) \approx -\frac{R_\infty}{n^2}\left[1 + \left(\frac{\alpha}{n}\right)^2\left(\frac{n}{J + \frac{1}{2}} - \frac{3}{4}\right) + \cdots\right] \tag{3}$$

(first derived by Sommerfeld on the basis of a relativistic generalization of Bohr's quasi-classical model). R_∞ is the Rydberg energy unit defined earlier, the subscript specifically referring to the case of an infinitely massive (and therefore fixed) nuclear center. The fine structure (proportional to α^2) can be interpreted as arising from two contributions: (a) the relativistic increase in electron mass—specifically the first-order correction $p^4/(8m^3c^2)$ to the kinetic energy $p^2/2m$ (with linear momentum $p = mv)^4$; and (b) the classical interaction between the magnetic dipole moment of the electron and the magnetic field produced at the electron location by the proton (seen as a circulating positive current from the rest frame of the electron). Explicit calculation of the second, or spin-orbit, contribution requires some caution, for a straightforward evaluation of the magnetic dipole energy expression fails to account for a needed factor of $\frac{1}{2}$ arising from a relativistic kinematic effect known as the Thomas precession.[5] In any event, substitution into Eq. (3) of the two allowed values $J = L \pm \frac{1}{2}$ (for $L \neq 0$) leads to fine structure energies of the form

$$E_{fs} = \zeta(nL)\begin{cases} L \\ -(L+1) \end{cases} \quad \text{for} \quad \begin{cases} J = L + \frac{1}{2} \\ J = L - \frac{1}{2} \end{cases} \tag{4}$$

in which the level of lower J lies lowest as illustrated in figure 1.2. The common factor $\zeta(nL)$ is a radial integral proportional to the expectation value of r^{-3}.[6]

A particularly significant feature of the Dirac solution is that the energy of any given state depends only on the quantum numbers n, J—and not explicitly on the orbital quantum number L. Thus, two states of the same electronic manifold such as $n^2S_{1/2}$ and $n^2P_{1/2}$ or $n^2P_{3/2}$ and $n^2D_{3/2}$ are predicted by Dirac's theory to have the same energy. It is perhaps not surprising that the energy of the atomic states depends on the magnitude of the total angular momentum J, since the electron is moving in a purely central electric field. The degeneracy is broken, however, by

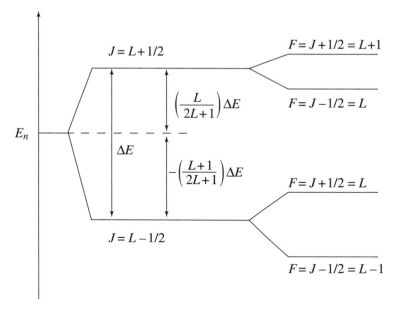

Figure 1.2 Normal ordering of hydrogen fine and hyperfine structure levels. For given L, the level with larger J lies higher; for given L and J, the level with larger F lies higher. The hyperfine splitting follows the same pattern as that shown for the fine structure splitting with the substitutions $S \to I$, $L \to J$, $J \to F$.

the introduction of any anisotropy such as would result from the presence of other electrons or external fields (neither of which is an issue here), or from the magnetic properties of the nucleus, which *is* a matter of concern.

The second term in Eq. (1),

$$E_{\text{hf}}(nLJF) = \frac{\alpha^2 g_p R_\infty}{n^3} \frac{m}{m_p} \frac{F(F+1) - J(J+1) - \frac{3}{4}}{J(J+1)(2L+1)}, \quad (5)$$

is the hyperfine energy arising from the magnetic dipole interaction between the electron and nuclear magnetic moments. [For a nucleus of charge Z, the right hand side of Eq. (5) is multiplied by Z^3.] The possibility of such a "spin-spin" interaction was first proposed around 1924 by Pauli. Now the proton is no longer just a center of attraction, but a particle with mass m_p, spin angular momentum quantum number $I = \frac{1}{2}$ [with the product $I(I+1)$ yielding the term $\frac{3}{4}$ in Eq. (5)], and gyromagnetic ratio $g_p \sim 2.8$. The gyromagnetic ratio is the proportionality factor relating the magnetic moment of a particle to its spin angular momentum in units of the appropriate Bohr magneton. Thus, the electron and proton magnetic moments can be expressed, respectively, as $g_e \mu_e S$ and $g_p \mu_p I$, in which the magnetons of the two particles

$$\mu_e = \frac{e\hbar}{2mc}, \qquad \mu_p = \frac{e\hbar}{2m_p c} \tag{6}$$

differ in magnitude by a factor of nearly 1840 (the ratio of proton to electron mass). In hydrogen, therefore, the hyperfine splitting is indeed much smaller than the separation between fine structure components, a situation that does not pertain to positronium ($e^+ e^-$), for example, in which the two bound particles have identical mass. Dirac's relativistic theory predicts an electron gyromagnetic ratio of precisely $g_e = 2$, which is twice the value expected from the classical dynamics of an orbiting particle with no internal structure. The departure of g_p from 2 is indicative of an internal distribution of charge, confirming that the proton is not an elementary particle.

The total angular momentum quantum number F assumes two values, $F = J \pm \frac{1}{2}$, for a nucleus with $I = \frac{1}{2}$. Thus, each hydrogen fine structure term $n^2 L_J$ is split into two hyperfine levels (see figure 1.2), with the level of lower F lying lowest, as in the case of spin-orbit coupling. A curious coincidence, by no means evident from the form of Eq. (5), is that the equation is valid for both $L = 0$ and $L \neq 0$. This is interesting because the equation is actually the end result of a calculation involving the product of the expectation value $\langle 1/r^3 \rangle$ (where r is the electron radial coordinate) and an angular factor that is proportional to L. Since $\langle 1/r^3 \rangle$ is itself proportional to L^{-1}, the entire expression becomes an indeterminate $0/0$ for S states and is consequently not valid. Physically, the difficulty arises from the fact that only for states with $L = 0$ is there a nonvanishing probability that the electron wavefunction overlaps the nucleus; r may be 0 (for a point nucleus), but there is still a finite "contact" interaction. The correct derivation of this contact term was provided by Fermi,[7] and Eq. (5) reduces precisely to Fermi's result when L is set to 0. Examination of Eq. (5) shows, therefore, that the hyperfine splitting is greatest in S states and diminishes with increasing L.

Although we are now considering a proton with finite mass, its motion to this point has still been neglected as evidenced by the fact that the energy unit of Eqs. (3) and (5) remains R_∞. The third term on the right hand side of Eq. (1), E_N, is an energy correction arising from the finite mass and size of the nucleus. To lowest order of approximation, one can take account of the fact that the electron and proton orbit a common center of mass by replacing the electron mass m in the Rydberg R_∞ by the reduced mass m_r,

$$m_r = \frac{mm_p}{m + m_p},$$ (7a)

thereby leading to a redefinition of the Rydberg constant for finite-mass nucleus:

$$R = \frac{R_\infty}{1 + \dfrac{m}{m_p}}.$$ (7b)

Underlying the replacement of R_∞ by R is the factorization of the electron wavefunction into a function of the coordinates of the center of mass and a function of the "reduced" coordinates (which approximate the electron coordinates to the extent that the ratio $m/m_p \to 0$). The difficulty with this approach, however, is that such a separation can be performed rigorously only for a nonrelativistic equation of motion (e.g., either Newton's or Schrödinger's), but is less meaningful in the context of special relativity.

Electron bound state energies are also affected by the finite size of the nucleus within which the electrostatic potential no longer varies strictly as $1/r$. In general, this serves to weaken the binding of states (principally S states) with wavefunctions that overlap the nucleus. Effects (which depend on m/m_p) arising from nuclear motion and those (which depend on $Z\alpha$) associated with relativity and electric charge are not separable and have been treated to desired degrees of approximation by various power-series expansions in both these small parameters.

The fourth term on the right hand side of Eq. (1), E_{QED}, represents radiative or quantum electrodynamic corrections arising out of virtual interactions of the bound electron with the vacuum field. These contributions are small but of considerable importance in the analysis of electric resonance spectra where one is concerned with energy differences

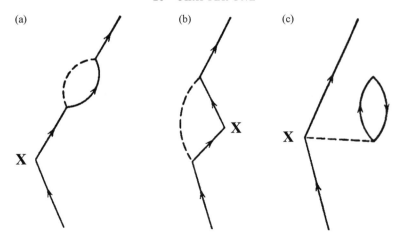

Figure 1.3 Feynman diagrams for the principal QED contributions to the energy of a hydrogen atom. A full line represents an electron (upward arrow) or positron (downward arrow); a broken line represents a photon. Each vertex marked by "x" signifies a scattering of the electron by the coulomb field of the nucleus. Diagrams (a) and (b) (and a third, not shown, in which absorption and emission of the virtual photon occur before electron scattering by the nucleus) contribute to the self-energy of the electron. Diagram (c), in which the electron is scattered by a photon produced from a virtual electron-positron pair, contributes to polarization of the vacuum.

between states in the same electronic manifold. In particular, the relative displacement of states of the same nJ quantum numbers is entirely due to QED and can be larger than hyperfine splittings. Mathematical expressions for these corrections—representative of an open-ended asymptotic series of ever more complicated virtual emissions and absorptions of photons or creations and annihilations of electron-positron pairs—are quite cumbersome and will not be needed here. It is useful, however, to examine briefly the physical significance of the simplest and most significant such virtual processes, illustrated in figure 1.3.

In the case of atomic hydrogen the largest contribution to the relative displacement of otherwise degenerate Dirac energy levels is the so-called electron self-energy term, portrayed diagrammatically by the emission and reabsorption of a single virtual photon. (A virtual process is one that can violate the conservation of energy, so long as it occurs over an undetectably small interval of time.) From a nonrelativistic, partly classical perspective, one can think of the electron as moving about in a sort of Brownian motion in response to random fluctuations of the zero-point radiation field in otherwise empty space. Although there is no net progress in any particular direction, the mean-square fluctuation in electron position does not vanish, and, as a result, the instantaneous potential that the electron experiences differs from the coulomb potential, $V(r) = -e^2/r$, by a small term proportional to $\nabla^2 V(r)$. For a potential varying as $1/r$, the Laplacian ∇^2 produces a point charge density at the origin [that is, a term proportional to the Dirac delta function $\delta(r)$] which, like a finite nuclear size, weakens the binding of S states, or in other words shifts the S levels "upward" on the hydrogen energy level diagram.[8]

A second (and, in hydrogen, smaller) QED contribution is that attributable to "polarization of the vacuum," represented diagrammatically by the virtual creation of a single electron-positron pair near the nucleus, the annihilation of this pair with subsequent creation of a photon, and the absorption of this photon by the bound electron. The fluctuating presence of electron and positron pairs serves to shield the "bare" charge $+e_0$ of the proton, in effect reducing it to the experimentally observed value $+e$ beyond a certain distance of the order of the Compton wavelength of the electron, $\lambda = h/mc \sim 2.4 \times 10^{-10}$ cm. The vacuum therefore behaves somewhat like a polarizable dielectric. Vacuum polarization shifts the energy level "downward" to the extent that the electron wavefunction penetrates the shielding and experiences a stronger proton charge.

In comparison to the separation between electronic manifolds represented by the Rydberg, which can be written as $R_\infty = \frac{1}{2}\alpha^2 mc^2$, the leading contributions to the fine structure splitting, hyperfine structure splitting, and QED splitting (of degenerate Dirac states) are respectively $\Delta E_{\text{fs}} \sim \alpha^4 mc^2$, $\Delta E_{\text{hf}} \sim \alpha^4 mc^2 (m/m_p)$, and $\Delta E_{\text{QED}} \sim \alpha^5 mc^2$. Evaluation of higher-order QED terms required for ever more stringent comparison with experiment is a difficult and tedious operation that has long ago surpassed the capacity of theorists to execute by hand without the aid of computers (not only for numerical evaluation of difficult integrals, but also for organization and symbolic manipulation of Feynman diagrams). However, the stunning agreement between the predicted and observed structures of the hydrogen atom is one of the seminal achievements of theoretical physics in the twentieth century.

1.2 SHAPES AND WIDTHS

Although the energy levels—and therefore energy intervals—of an isolated atom decoupled from its environment are the eigenvalues of a
well-defined hamiltonian and can in principle be calculated with arbitrary accuracy, optical transitions between states of real atoms do not
occur with correspondingly infinite sharpness. What ideally would be a
"line" (in a spectroscope) resulting from spontaneous radiative decay
is actually a broadened spectral lineshape as a result of three principal
types of interactions.

The first derives from the finite natural lifetime τ of an excited atomic
state. This does not mean that all atoms in the specified state decay in τ
seconds; rather, it is a statistical measure of the mean duration of a large
number of such states. According to the Heisenberg uncertainty principle
applied to time and energy, a finite lifetime τ implies a corresponding
energy uncertainty $\Delta E \geq \hbar/\tau$ and therefore a natural level width $\Delta\omega \sim \tau$
(expressed in terms of angular frequency $\omega = E/\hbar$). From a semiclassical
perspective, the radiating atom is analogous to a charged oscillator whose
train of radiation is terminated when the atom returns to its ground
state. Basic Fourier analysis shows that an interrupted wave train cannot
be monochromatic.

According to quantum mechanics the probability amplitude $a(t)$ for
a (two-level) atom excited at time $t = 0$ to be found in the excited
state at time t later is $a(t) = e^{-i\omega_0 t}e^{-\gamma_r t}$. Here ω_0 is the Bohr frequency
for transition to the ground state, and $\gamma_r = 1/\tau$ is the radiative decay
rate. The Fourier transform of this amplitude, which gives in effect the
contribution of each frequency component in the emitted radiation, is

$$f(\omega) = \int_0^\infty a(t)e^{i\omega t}\, dt = \frac{i}{\omega - \omega_0 + i\gamma_r}, \qquad (8a)$$

from which follows the Lorentzian spectral profile

$$I(\omega) = \text{constant} \times |f(\omega)|^2 = \frac{\gamma/\pi}{(\omega - \omega_0)^2 + \gamma_r^2}, \qquad (8b)$$

where the constant inserted into Eq. (8b) normalizes the lineshape to
unit area: $\int_0^\infty I(\omega)\, d\omega = 1$. In other words, $I(\omega)\, d\omega$ is the fraction of
light intensity emitted in the frequency range ω to $\omega + d\omega$.

The width $\Delta\omega$ of a spectral line is often defined as the "full width at half maximum," that is, the frequency interval between the two points at which $I(\omega)$ is one-half its maximum value. It then follows readily from Eq. (8b) that $\Delta\omega = \gamma_r$, confirming quantitatively the result previously obtained by heuristic application of the uncertainty principle. In a fully quantum mechanical treatment of radiative decay, the natural linewidth is a consequence of the interaction of an excited atom with the quantized vacuum electromagnetic field (a process that will be taken up in detail in chapter 5).

A second cause of broadening results from elastic collisions of an excited atom with other atoms, ions, or free electrons. Although such encounters do not affect the energy of the radiating atom, they interrupt the phase of the emitted wave train at random intervals, thereby giving rise to a spectrum of frequencies about the Bohr frequency ω_0. A wave train, persisting for a time interval $0 \leq t' \leq t$ until terminated by a collision, has the Fourier components

$$f(\omega; t) = \int_0^t e^{-i\omega_0 t'} e^{i\omega t'}\, dt' = \frac{e^{i(\omega - \omega_0)t} - 1}{i(\omega - \omega_0)}. \qquad (9)$$

For independent random collisions occurring at the rate of γ_c per second, the radiation intensity at t [proportional to $|f(\omega; t)|^2$] must be weighted by the probability (derived from a Poisson distribution)

$$P(t)\, dt = \gamma_c e^{-\gamma_c t}\, dt \qquad (10)$$

that no collision occurs in the interval t and then one collision occurs in the interval between t and $t + dt$. The resulting lineshape

$$I(\omega) = \text{constant} \times \int_0^\infty P(t)|f(\omega; t)|^2\, dt = \frac{\gamma_c/\pi}{(\omega - \omega_0)^2 + \gamma_c^2} \qquad (11)$$

is again of Lorentzian form, but with linewidth γ_c. To good approximation, $\gamma_c = \rho\sigma\bar{v}$, where $\sigma \sim \pi r^2$ is the cross section of the excited atom (of effective radius r), ρ is the number density of excited atoms, and $\bar{v}$ is the relative mean speed of colliding particles.

The third, and ordinarily most significant, contribution to lineshape broadening arises from the thermal motion of radiating atoms, the velocities of which are distributed according to a Maxwellian distribution

$$\frac{dN}{N} \propto e^{-Mv_x^2/2kT} \, dv_x. \tag{12}$$

Here dN/N is the fraction of atoms of mass M with x-component of velocity between v_x and $v_x + dv_x$; k is Boltzmann's constant and T is the absolute temperature. Since each atomic light source is moving with respect to the stationary observer (or detecting device), the perceived frequency ω is Doppler-shifted from the resonance frequency ω_0 in the atomic rest frame according to

$$\omega = \gamma\omega_0\left(1 - \frac{v}{c}\cos\theta\right), \tag{13}$$

where θ is the angle between the atomic beam and line of sight. (Do not confuse the relativistic factor γ with a decay rate.) For $\theta = 0$, the detected radiation is antiparallel to the beam and maximally "red-shifted"; for light received at right angles to the beam, the Doppler shift is second order in v/c and ordinarily unimportant except for highly relativistic beams.

As a result of the velocity distribution (12) and Doppler shift (13), moving atoms give rise to a Gaussian emission lineshape

$$I(\omega) = \frac{e^{-[(\omega-\omega_0)/\gamma_D]^2}}{\sqrt{\pi}\gamma_D} \qquad \left(\gamma_D = \sqrt{\frac{2kT}{Mc^2}}\,\omega_0\right) \tag{14}$$

with full width at half maximum $\Delta\omega = \sqrt{\ln 2}\,\gamma_D$. A comparison of Lorentzian and Gaussian profiles is shown in figure 1.4.

Spectroscopic investigation of the fine structure of hydrogen began around 1887 with the work of Michelson and Morley,[9] who first discovered by optical interferometry that the Balmer α line ($n = 3$ to $n = 2$ transitions at wavelength $\lambda \sim 656.3$ nm) was actually a doublet. Since that time and until the introduction of various methods of nonlinear laser spectroscopy in the 1970s, the efforts to obtain high-resolution hydrogen spectra by optical spectroscopy have been numerous and skillful, but inevitably thwarted by Doppler broadening. At room temperature (300 K), thermal motion of hydrogen atoms broadens the Balmer α line by approximately 5.7 GHz, a width considerably greater than all the fine structure splittings in the $n = 3$ manifold.

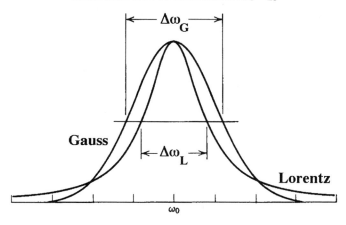

Figure 1.4 Lorentzian and Gaussian lineshapes (both normalized to unit area).

It follows from Eq. (14) that there are effectively two experimental ways to reduce Doppler broadening: One can either lower the temperature of the atoms or detect radiation of lower frequency (or both). Since the Doppler shift is proportional to frequency, Doppler broadening poses a more serious limitation in the optical domain ($\sim 10^{14}$–10^{15} Hz) than in the domain of radio and microwave frequencies (roughly 10^6–10^{10} Hz).

Except for transitions between levels of very high principal quantum number (Rydberg states), radiowaves are ordinarily produced or absorbed by transitions occurring *within* an electronic manifold. That atomic selection rules permitted $\Delta n = 0$ transitions was first realized by W. Grotrian in 1928, but radiation sources with which to perform such experiments were not developed until the Second World War brought a pressing need for radar. As discussed in the preface, Willis Lamb and his group carried out the first investigations (~ 1947) of hydrogen fine structure by a resonant microwave method. These experiments employed a thermal atomic beam with H atoms excited into the metastable $2S$ state (lifetime $\sim 1/7$ second) by electron bombardment.

It is important to note that Doppler broadening would be eliminated entirely if all atoms in a beam were moving at the same velocity. There would then be a single atomic rest frame in which the effective temperature of the atoms is $T = 0$ K. If atomic velocity is sharply defined, the Doppler shift of absorbed or emitted radiation constitutes a simple correction, rather than a limitation on the precision with which spectral lines can be measured. The use of an "accelerated" atomic beam—that is, a beam of neutral atoms produced by charge-exchange collisions of electrostatically accelerated ions (protons in the case of hydrogen) with effectively stationary target atoms—provides a close approximation to this ideal condition. Also, by rapidly transporting excited atoms out of the target chamber to highly evacuated chambers for radiofrequency spectroscopy and optical detection, a fast atomic beam avoids pressure broadening as well as Doppler broadening, and permits the study of many relatively short-lived states.

Although a spectroscopist may have some control over Doppler and pressure broadening, the natural lifetime is generally independent of external conditions.[10] Nevertheless, the effect of spontaneous decay on linewidth can differ significantly between optical and rf spectroscopy. For example, in a theoretical study of spectral lines in gases[11] undertaken (~ 1933) as part of his doctoral thesis work with Eugene Wigner, Weisskopf demonstrated that the width of an optical transition between two decaying states is the *sum* of the individual decay rates. In transitions induced between two coupled states by an external rf field, however, the transition probability depends on the *difference* of the decay rates (as will be shown in succeeding chapters). Thus, the rf resonance profile arising from two coupled transient states of equal lifetime has the same width as if the states were infinitely long lived (although of course the resonance amplitude will be smaller, since exponential decay reduces the number of atoms passing through the field). The actual shape of the rf profile is neither Lorentzian nor Gaussian, but a superposition of trigonometric (or hyperbolic trigonometric) functions; the width of the profile depends not only on state lifetimes, but also on the electric dipole matrix element—in effect the rf power—coupling the states.

The combined use of a fast beam and several rf fields makes possible various procedures for narrowing lineshapes. First, in cases of overlapping transitions, it may be feasible to remove or "quench" a certain population of atoms by driving them into short-lived states that decay before (or after) the remaining atoms of interest are probed by the rf "spectroscopy" field. Second, the use of two spatially separated, coherently oscillating rf fields ("Ramsay" fields) gives rise to lineshapes characteristic of a two-slit diffraction-interference pattern. The central fringe is narrower than the profile obtained by use of a single rf field. More-over, the greater the separation of the two coherently oscillating fields, the longer is the average lifetime of the subpopulation of excited atoms that traverses the intervening field-free region—and hence (by the un-certainty principle) the more sharply defined is the energy of the state, and the narrower is the resulting lineshape.

These and other spectroscopic features of coupling decaying atomic states by rf electric fields will be explored in the following chapters. So precise are the measurements of atomic structure provided by rf meth-ods that only now, some forty years after the first laser, is optical spec-troscopy catching up.

NOTES

1. A bound electron-positron system (positronium) is simpler than hydrogen since the positron, unlike the proton, has no internal structure and is not subject to strong nuclear interactions. As far as is known, however, positronium does not occur naturally.

2. See, for example, M. P. Silverman, *And Yet It Moves: Strange Systems and Subtle Questions in Physics* (Cambridge University Press, New York, 1993), chap. 2, "Quantum Beats and Giant Atoms."

3. H. A. Bethe and E. E. Salpeter, *Quantum Mechanics of One- and Two-Electron Atoms* (Springer, Heidelberg, 1957), p. 68.

4. The exact relativistic expression for kinetic energy is $K = mc^2(\gamma - 1) = mc^2(\sqrt{1 + (p/mc)^2} - 1)$ which when expanded in a power series in p/mc leads to the term in the text.

5. L. H. Thomas, *Nature* **117** (1926):514. The basic idea underlying the Thomas correction is that two successive Lorentz transformations do not in general commute, but instead are equivalent to a Lorentz transformation plus a three-dimensional rotation. As a result, keeping track correctly of the pro-ton motion seen from the electron rest frame requires adding to the magnetic dipole equation of motion a term equivalent to a precession.

6. $\zeta(nL)$ is the radial matrix element

$$\left\langle nL \left| \frac{\hbar^2}{2m^2c^2} \frac{1}{r} \frac{\partial V}{\partial r} \right| nL \right\rangle$$

where, for hydrogenic atoms, the coulomb potential seen by the electron is $V(r) = -Ze^2/r$.

7. E. Fermi, Z. *Physik* **60** (1930):320.

8. T. A. Welton, "Some Observable Effects of the Quantum-Mechanical Fluctuations of the Electromagnetic Field," *Phys. Rev.* **74** (1948):1157.

9. A. A. Michelson and E. W. Morley, *Phil. Mag.* **74** (1887):466.

10. The rate of decay of an atomic state depends on the number of modes of the vacuum electromagnetic field into which the decay can occur. This number can be reduced, and the decay inhibited, if an atom is confined to an effectively two-dimensional cavity—that is, a cavity of height comparable to the radiation wavelength.

11. V. Weisskopf, "The Width of Spectral Lines in Gases," translated from *Physikalische Zeitschrift* **34** (1933) 1 and reprinted (abridged) in W. R. Hindmarsh, *Atomic Spectra* (Pergamon, Oxford, 1967), pp. 328–363.

CHAPTER 2

The Driven Two-Level Atom

2.1 DYNAMICS OF A TWO-LEVEL ATOM

Experimental progress in the determination of atomic structure may very well be regarded as the attempt to achieve the ideal of spectroscopic purity: transitions from a single quantum state to another single quantum state driven by a completely polarized monochromatic radiation field. From a theoretical perspective, however, the analysis of a two-state quantum system interacting with an oscillating field leads to coupled differential equations for which *no* exact closed-form solution has been found, even in the case where the field is treated classically.

The simplest approximate solution to this problem is obtained by reducing the time dependence to that of a rotating field. In the rotating-wave approximation (RWA), the condition of resonance—that is, the maximum probability for transition out of an initially populated state or into an initially empty state—occurs precisely at the Bohr frequency ω_0 of the atom. This may seem reasonable, but systems driven by an external oscillating field do not in fact behave exactly this way. The first consideration of the effects of neglecting the "antiresonant" or "counter-rotating" component of the oscillating field was given by Bloch and Siegert[1] for the case of magnetic resonance between two stable spin-1/2 states. Their correction to the RWA theory yielded a small shift in the resonance maximum proportional to the square of the magnetic field and inversely proportional to the level separation. Thus, the correction can be fairly significant for close-lying levels driven by a strong field.

The Bloch-Siegert result, however, is not a priori applicable to transitions driven between rapidly decaying quantum states. Also, quite apart from the issue that has traditionally interested spectroscopists most—namely, the relationship between the true resonance frequency and the location of the peak of a spectral line—there is a more general issue of being able to determine in a satisfactory way, without resorting each time to numerical integration, the effect of the applied field on the coupled states; in particular, to be able to predict the full resonance line-shape. To an excellent approximation, this information is provided by the oscillating-field theory (OFT) developed in this chapter.[2]

Let us start with a quantum system comprising two quasi-stationary states $|1\rangle$ and $|2\rangle$ interacting in the electric dipole approximation with a classical oscillating field $\mathbf{E}_0 \cos \omega t$. These atomic states are quasi-stationary in the sense that, although they are energy eigenfunctions of a Hermitian Hamiltonian,[3] their interaction with the vacuum electro-magnetic field results in a finite lifetime and hence in a natural level width. The dynamics of the atom–rf field system can be described by the Hamiltonian

$$\mathcal{H} = \mathcal{H}_0 + \mathcal{H}_D - \boldsymbol{\mu}_E \cdot \mathbf{E}_0 \cos \omega t \tag{1}$$

in which $\mathcal{H}_0$ is a Hermitian operator yielding the energy levels through the eigenvalue equation

$$\mathcal{H}_0 |j\rangle = \hbar \omega_j |j\rangle \qquad (j = 1, 2), \tag{2}$$

and $\mathcal{H}_D$ is a phenomenological anti-Hermitian operator accounting for the level decay rates through the analogous relation

$$\mathcal{H}_D |j\rangle = -\tfrac{1}{2} i \hbar \gamma_j |j\rangle \qquad (j = 1, 2). \tag{3}$$

The electric dipole operator $\boldsymbol{\mu}_E$ is simply equal to $-e\mathbf{r}$ in the case of a single bound electron (of charge $-e$) with coordinate operator $\mathbf{r}$. For well-defined states to exist, the level width must be much smaller than the energy eigenvalue; it will be assumed, therefore, that $\omega_j \gg \gamma_j$ for all states. The mean lifetime of a state is the reciprocal of the decay constant introduced in Eq. (3): $\tau_j = \gamma_j^{-1}$.

Throughout this book it will be convenient to express atomic energies and decay rates in units of angular frequency (equivalent to setting $\hbar = 1$). Thus, in the absence of coupling to an external driving field, the solution to the Schrödinger equation[4]

$$\mathcal{H} |\Psi\rangle = i \frac{d}{dt} |\Psi\rangle, \tag{4}$$

where $\mathscr{H}$ comprises only $\mathscr{H}_0 + \mathscr{H}_D$, takes the form

$$|\Psi(t)\rangle = \exp(-i\mathscr{H}t)|\Psi(0)\rangle, \tag{5}$$

thereby giving rise to amplitudes $\langle j \mid \Psi(t)\rangle$ with the simple time dependence $e^{-i\omega_j t}e^{-\frac{1}{2}\gamma_j t}$. The validity of presuming an exact exponential decay for each eigenstate [implied in Eq. (3)] is well entrenched in atomic physics and ordinarily justified by the quantum electrodynamic treatment of Weisskopf and Wigner referred to in the preface. Actually, the assumption is questionable for times either much longer or much shorter than the lifetime of the atomic state, but this will not be an issue in the following discussion. For the range of lifetimes of usual interest to spectroscopists, the assumption of exponential decay in vacuum is rigorously supported on both experimental and theoretical grounds. We will examine later in chapter 5 the influence of the rf field on radiative decay.

At time t_0 an atom enters the oscillating field. The state of the atom at a later time t can be represented by a superposition of its two quasi-stationary states,

$$|\Psi_t\rangle = c_1(t)|1\rangle + c_2(t)|2\rangle. \tag{6}$$

Substitution of Eq. (6) into Eq. (4), followed by projection upon each basis state, yields two coupled differential equations which can be readily cast into the form of a matrix equation,

$$\frac{d}{dt}\begin{pmatrix} c_1(t) \\ c_2(t) \end{pmatrix} = \begin{pmatrix} -(i\omega_1 + \frac{1}{2}\gamma_1) & -iV(e^{i\omega t} + e^{-i\omega t}) \\ -iV(e^{i\omega t} + e^{-i\omega t}) & -(i\omega_2 + \frac{1}{2}\gamma_2) \end{pmatrix}$$
$$\times \begin{pmatrix} c_1(t) \\ c_2(t) \end{pmatrix} = H\Psi, \tag{7}$$

where the interaction between the atom and the rf field is specified by the quantum mechanical matrix element

$$V = \frac{\langle 1| - \boldsymbol{\mu}_E \cdot \mathbf{E}_0|2\rangle}{2\hbar}. \tag{8}$$

To keep the analysis from becoming encumbered by more notation than is necessary, the same symbol Ψ will label both the state vector and the representation as a column vector of amplitudes. The matrix H defined in Eq. (7), whose elements have the dimension of angular frequency, corresponds to $-i\mathcal{H}$; for want of a better term, we shall continue to refer to H (and matrices obtained by various transformations of H) as a Hamiltonian.

To eliminate from the time dependence of Eq. (7) the high frequencies deriving from the Hamiltonian $\mathcal{H}_0$, in order that the resulting equation evolve at the slower frequency of the applied field (which is primarily the interaction of interest here), it is useful to reexpress Eq. (7) in the interaction representation defined by the transformation

$$\Psi_I = \begin{pmatrix} e^{i\omega_1 t} & 0 \\ 0 & e^{i\omega_2 t} \end{pmatrix} \Psi \equiv U_I \Psi. \tag{9}$$

This leads to the equation

$$\frac{d}{dt}\Psi_I = \left[U_I H U_I^\dagger + U_I \frac{dU_I^\dagger}{dt} \right] \Psi_I \equiv H_I \Psi_I, \tag{10}$$

whose explicit form

$$\frac{d}{dt}\Psi_I = \begin{pmatrix} -\frac{1}{2}\gamma_1 & -iV(e^{-i\Omega t} + e^{i\Omega' t}) \\ -iV(e^{i\Omega t} + e^{-i\Omega' t}) & -\frac{1}{2}\gamma_2 \end{pmatrix} \Psi_I \tag{11}$$

contains "resonant" or "rotating" terms at the frequency

$$\Omega = \omega - \omega_0 \tag{12a}$$

and "anti-resonant" or "counter-rotating" terms at the frequency

$$\Omega' = \omega + \omega_0, \tag{12b}$$

where

$$\omega_0 = \omega_1 - \omega_2 \tag{12c}$$

is the Bohr frequency for the two-level system. It it assumed in this section that $\omega_1 > \omega_2$—that is, state $|1\rangle$ lies higher than state $|2\rangle$—so that ω_0 is a positive number and Ω vanishes at the condition of resonance $\omega = \omega_0$; were the level ordering opposite, then Ω' would become the frequency of the resonant terms. A generalization of definitions (12a, b) will be made in a later chapter concerned with rf coupling of more than two states.

The RWA solution of Eq. (11) (and hence of the original Schrödinger equation) consists in neglecting all terms containing Ω'. Before proceeding to the OFT solution, which yields results almost indistinguishable from numerical integration of the Schrödinger equation, let us first examine the RWA solution further.

2.2 ROTATING-WAVE APPROXIMATION

If terms containing the antiresonant frequency Ω' are dropped from Eq. (11), it is then possible to eliminate all time dependence from the coefficient matrix (that is, the Hamiltonian) of the Schrödinger equation by making the transformation

$$\Psi_R = \begin{pmatrix} e^{i\Omega t/2} & 0 \\ 0 & e^{-i\Omega t/2} \end{pmatrix} \Psi_I \equiv U_R \Psi_I. \tag{13}$$

This leads to the equation

$$\frac{d}{dt}\Psi_R = \left[U_R H_I U_R^\dagger - U_R \frac{dU_R^\dagger}{dt} \right]\Psi_R \equiv H_R \Psi_R, \tag{14}$$

in which all elements of the transformed Hamiltonian

$$H_R = \begin{pmatrix} -\frac{1}{2}(\gamma_1 - i\Omega) & -iV \\ -iV & -\frac{1}{2}(\gamma_2 + i\Omega) \end{pmatrix} \tag{15}$$

are constants for a given magnitude and frequency of the applied field.

Once a Schrödinger equation (for any number of coupled states) has been reduced to the basic form $d\Psi/dt = H\Psi$ where H is a constant matrix, the solution may be obtained immediately by a standard procedure.[5] First, diagonalize the matrix H by solving the eigenvalue problem

$$HX_i = \varepsilon_i X_i, \tag{16}$$

which yields a set of n eigenvalues and eigenvectors for a matrix of n dimensions. Let us assume the eigenvalues are all distinct, for the spectroscopic problems to be treated in this book fall into that category. Second, arrange the eigenvalues to form the matrix ε with elements ε_i along the principal diagonal (and zeroes everywhere else); likewise, arrange the eigenvectors in the same order so as to create the matrix

$$C = X_1 X_2 \cdots X_n \tag{17}$$

that transforms H into ε according to

$$H = C\varepsilon C^{-1} \tag{18}$$

(known as a similarity transformation). The solution to the Schrödinger equation then takes the form

$$\Psi(t) = e^{Ht}\Psi(0) = Ce^{\varepsilon t}C^{-1}\Psi(0), \tag{19}$$

in which $e^{\varepsilon t}$ is a diagonal matrix with elements $e^{\varepsilon_j t}$ ($j = 1, \ldots, n$).

In the case of a two-state system with time-independent Hamiltonian matrix H (with elements H_{ij}), application of the procedure leading to Eq. (19) results in the symmetric expression

$$\Psi(t) = e^{(H_{11} + H_{22})t/2}$$

$$\times \begin{pmatrix} \cosh \bar{\omega}t + \dfrac{(H_{11} + H_{22})}{2\bar{\omega}} \sinh \bar{\omega}t & \dfrac{H_{12}}{\bar{\omega}} \sinh \bar{\omega}t \\[3mm] \dfrac{H_{21}}{\bar{\omega}} \sinh \bar{\omega}t & \cosh \bar{\omega}t + \dfrac{(H_{22} - H_{11})}{2\bar{\omega}} \sinh \bar{\omega}t \end{pmatrix} \Psi(0),$$

$$\tag{20}$$

where

$$\bar{\omega} = \sqrt{\tfrac{1}{4}(H_{11} - H_{22})^2 + H_{12}H_{21}}. \tag{21}$$

Since the model of a two-state system arises frequently in physics—and not just in the context of atomic spectroscopy—it is worth noting that (for a two-dimensional system only) the equation of motion can be solved more expeditiously and with greater physical insight by an approach different from the foregoing standard procedure. This more elegant method (which I have employed in previous books concerned with quantum physics[5] and physical optics[6]) begins by expanding the Hamiltonian (or equivalent matrix) in a basis consisting of the unit 2×2 matrix σ_0 and the three Pauli spin matrices σ_j ($j = 1, 2, 3$). From the defining algebraic properties of the Pauli matrices

$$\sigma_j \sigma_k = i \sum_{\ell=1}^{3} \varepsilon_{jk\ell} \sigma_\ell + \delta_{jk} \sigma_0 \tag{22}$$

[where $\varepsilon_{jk\ell}$ is the completely antisymmetric tensor (or Levi-Cività symbol) and δ_{jk} is the Kronecker delta symbol[7]], one can integrate $d\Psi/dt = H\Psi$ directly without first having to solve an eigenvalue problem. Moreover, the solution contains linear combinations of the elements H_{ij} that have a direct physical significance depending on the nature of the system being modeled. For example, the combinations of H_{ij} may be components of the polarization vector of a spin-1/2 particle, or the Stokes' parameters of a light beam, or the angular velocity vector of a precessing system. Also, apart from their transparent physical interpretation, certain combinations can be identified with the mathematical invariants of the two-dimensional matrix H, such as the trace [to which the global phase in Eq. (20) is proportional] and the determinant [which is proportional to $\bar{\omega}$ defined in Eq. (21)]. It is often useful to have a solution in a manifestly invariant form so that one can ascertain easily how physical quantities derived from the solution depend on changes of coordinate system or representation.

In any event, elegance aside, Eq. (20) expresses the wavefunction of an arbitrary two-state system interacting with a time-independent field in a form that can be directly applied to the decaying two-level atom. In this case, after retracing all the transformations back to the original

Schrödinger equation with initial conditions $t_0 = 0$ and $\Psi(0) = \begin{pmatrix} c_1^0 \\ c_2^0 \end{pmatrix}$, one obtains the RWA wavefunction

$$\begin{pmatrix} c_1(t) \\ c_2(t) \end{pmatrix} = \begin{pmatrix} e^{-(i\omega_1 + \frac{1}{2}\gamma_1)t} & 0 \\ 0 & e^{-(i\omega_2 + \frac{1}{2}\gamma_2)t} \end{pmatrix} \begin{pmatrix} I_{11} & I_{12} \\ I_{21} & I_{22} \end{pmatrix} \begin{pmatrix} c_1^0 \\ c_2^0 \end{pmatrix} \tag{23}$$

in a form separating the time evolution induced by $\mathcal{H}_0 + \mathcal{H}_D$ from the effects of the applied field embodied in the elements

$$I_{11} = e^{(\Gamma - i\Omega)t/2} \left[\cosh \nu t - \left(\frac{\Gamma - i\Omega}{2\nu} \right) \sinh \nu t \right], \tag{24a}$$

$$I_{12} = -ie^{(\Gamma - i\Omega)t/2} \left(\frac{V}{\nu} \right) \sinh \nu t, \tag{24b}$$

$$I_{21} = -ie^{-(\Gamma - i\Omega)t/2} \left(\frac{V}{\nu} \right) \sinh \nu t, \tag{24c}$$

$$I_{22} = e^{-(\Gamma - i\Omega)t/2} \left[\cosh \nu t + \left(\frac{\Gamma - i\Omega}{2\nu} \right) \sinh \nu t \right]. \tag{24d}$$

Only the difference in decay rates

$$\Gamma = \tfrac{1}{2}(\gamma_1 - \gamma_2), \tag{24e}$$

and not the individual decay constants, enters each interaction matrix element I_{ij}. Thus, the resonance lineshape for transitions induced betweeen decaying states of the same lifetime is identical to that for stable states apart from the diminution in intensity engendered by the first factor (decay matrix) in Eq. (23). The factor

$$\nu = \sqrt{\tfrac{1}{4}(\Gamma - i\Omega)^2 - V^2}, \tag{24f}$$

containing the mean decay rate difference Γ, detuning frequency Ω, and interaction matrix element V, is equivalent to $\bar{\omega}$ in Eq. (21) to within a factor of i. In the classical "vector model" of the coupling of two nondecaying stationary states, whereby the transition process is represented by a precessing dipole, ν is related to the angular frequency of precession. With $\Gamma = 0$ and $\nu = i\sqrt{\Omega^2 + V^2}$, the hyperbolic functions of Eqs. (24a–d) transform according to $\sinh(ix) = i\sin(x)$ and $\cosh(ix) = \cos(x)$, resulting in undamped periodic excursions of the atom between its two states. For unstable states, resonant ($\Omega = 0$) transitions are oscillatory only if $V > \frac{1}{2}\Gamma$.

Although the intermediate steps leading from Eq. (15) to the final solution (23) must be left to the literature,[2] several comments concerning the diagonalization of H_R are in order. The matrix that effects the similarity transformation $C_R H_R C_R^{-1}$ can be written as

$$C_R = \begin{pmatrix} 1 & -\kappa_R \\ \kappa_R & 1 \end{pmatrix}, \tag{25a}$$

in which the off-diagonal element κ_R, defined by

$$\kappa_R = \frac{\theta_R}{1 + \sqrt{1 + \theta_R^2}} \tag{25b}$$

with

$$\theta_R = \frac{2iV}{\Gamma - i\Omega}, \tag{25c}$$

vanishes with vanishing interaction matrix element V. In the limit that $\kappa_R \to 0$, C_R becomes the unit matrix; the two states are then uncoupled with $|1\rangle$ associated with eigenvalue ε_R^- and $|2\rangle$ with eigenvalue ε_R^+ where

$$\varepsilon_R^\pm = -\frac{1}{4}(\gamma_1 + \gamma_2) \pm \sqrt{\frac{1}{4}(\Gamma - i\Omega)^2 - V^2} \tag{26}$$

for arbitrary V.

The parameter κ_R may appear therefore as a small coupling constant giving the extent of contamination of one quasi-stationary state by the other. This is not strictly true, however, since κ_R need not be small even for small interaction strength V, particularly at resonance (see figure 2.1). For the case of resonant coupling of two states of the same lifetime (as in magnetic resonance), $|\kappa_R| = 1$; the effect on κ_R is the same as if $V \to \infty$. Under these circumstances the states are completely mixed.

We shall examine the characteristics of the lineshapes to which Eq. (23) gives rise later in conjunction with the lineshapes rendered by the OFT wavefunction. For the present it is worth noting from the form of the eigenvalues in Eq. (26) that the transition probabilities as a function of applied frequency lead to lineshapes symmetric about $\Omega = 0$ when

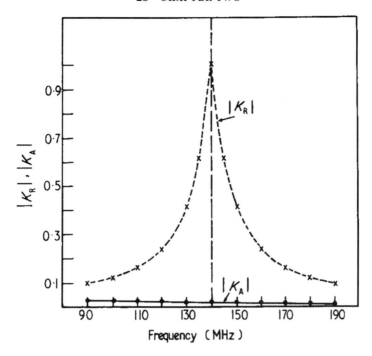

Figure 2.1 Comparison of RWA and OFT transformation parameters as a function of applied rf frequency for the conditions $\Gamma = 0$, $\omega_0 = 140$ MHz, $V = 5$ MHz.

one state is initially populated and the other unpopulated. Thus the frequency at which the peak or valley of a RWA lineshape occurs is the exact Bohr frequency of the atom.

2.3 OSCILLATING-FIELD THEORY

The OFT solution of the two-level decaying atom starts with the Schrödinger equation in the interaction representation, Eq. (11), and transforms—*not* to the reference frame of the rotating field—but rather to the reference frame of the antiresonant or counter-rotating field. This is accomplished by a matrix of the same form as Eq. (25a), but with $-\Omega'$ replacing Ω in Eq. (25c). The outcome is a Hamiltonian matrix which, as before, can be separated into a time-independent part and a time-dependent part. No part of the equation is dropped at this time, however. Instead, one follows the earlier procedure to diagonalize the time-independent matrix and then reexpress the entire differential equation in terms of the diagonalized basis. At this stage we are still in the reference frame of the counter-rotating field—but the primary secular effects of that field are incorporated in the eigenvalues

$$\varepsilon_A^{\pm} = -\tfrac{1}{4}(\gamma_1 + \gamma_2) \pm \sqrt{\tfrac{1}{4}(\Gamma + i\Omega')^2 - V^2} \qquad (27)$$

and transformation parameter

$$\kappa_A = \frac{\theta_A}{1 + \sqrt{1 + \theta_A^2}} \qquad (28a)$$

with

$$\theta_A = \frac{2iV}{\Gamma + i\Omega'}. \qquad (28b)$$

Next, the differential equation is transformed into the reference frame of the *rotating* field by a matrix transformation of the form of Eq. (13), but with ω replacing Ω. [Relative to the original laboratory frame of the Schrödinger equation, the transformation parameter to the rotating frame is $\Omega/2$ in one sense, whereas the transformation parameter to the counter-rotating frame is $-\Omega'/2$ in the opposite sense; thus, the transformation between the two frames is given by the parameter $\Omega/2 - (-\Omega'/2) = \omega$.] No approximations have yet been made. The resulting differential equation

$$\frac{d}{dt}\Psi_R = \left[\begin{pmatrix} \varepsilon_A^- + i\omega & \dfrac{-iV}{1 + \kappa_A^2} \\ \dfrac{-iV}{1 + \kappa_A^2} & \varepsilon_A^+ - i\omega \end{pmatrix} \right.$$
$$\left. + \begin{pmatrix} \dfrac{-2iV\kappa_A}{1 + \kappa_A^2}\cos 2\omega t & \dfrac{iV\kappa_A^2}{1 + \kappa_A^2}e^{4i\omega t} \\ \dfrac{iV\kappa_A^2}{1 + \kappa_A^2}e^{-4i\omega t} & \dfrac{2iV\kappa_A}{1 + \kappa_A^2}\cos 2\omega t \end{pmatrix} \right]\Psi_R \qquad (29)$$

is exact and contains all the information of the original Schrödinger equation.

The virtues of Eq. (29) lie in the following considerations. First, unlike the parameter κ_R, which increases in absolute magnitude as $\omega \to \omega_0$ and reaches its maximum at resonance, κ_A diminishes with increasing ω since it is roughly inversely proportional to $\omega + \omega_0$. In figure 2.1 is shown the variation in both transformation parameters as a function of applied frequency. The functions θ_A and κ_A are small to the extent that the resonance frequency and/or decay rate difference is larger than the interaction coupling the levels—a condition ordinarily applicable to electric resonance experiments on atomic fine structure levels (except for very highly excited atoms for which level separations are exceptionally small and natural lifetimes long). Second, not only is κ_A normally small, but it multiplies highly oscillatory factors in 2ω and 4ω in the second matrix term, which have negligible effect near the center of the lineshape.

To a good approximation, therefore, let us neglect the time-dependent matrix in Eq. (29) and retain only the time-independent part of the Hamiltonian. This time-independent part is diagonalized, and we proceed in precisely the same fashion as previously explained in the case of the RWA solution. From the resulting eigenvalues

$$
\varepsilon^{\pm} = -\tfrac{1}{4}(\gamma_1 + \gamma_2)
$$

$$
\pm \sqrt{\tfrac{1}{4}\{(\Gamma + i\Omega')(1 + \theta_A^2)^{1/2} - 2i\omega\}^2 - \frac{V^2}{(1 + \kappa_A^2)^2}} \qquad (30)
$$

and transformation parameter [in a matrix of the form of Eq. (25a)]

$$
\kappa = \frac{iV}{\varepsilon_A^{+} - \varepsilon^{-} - i\omega}, \qquad (31)
$$

we can construct, by retracing all transformations back to the original Schrödinger equation, the OFT solution

$$
\begin{pmatrix} c_1(t) \\ c_2(t) \end{pmatrix} = \begin{pmatrix} e^{-(i\omega_1 + \frac{1}{2}\gamma_1)t} & 0 \\ 0 & e^{-(i\omega_2 + \frac{1}{2}\gamma_2)t} \end{pmatrix}
$$

$$
\times \begin{pmatrix} J_{11} & J_{12} \\ J_{21} & J_{22} \end{pmatrix} \begin{pmatrix} c_1^0 \\ c_2^0 \end{pmatrix} \qquad (32)
$$

with interaction matrix elements

$$J_{11} = \frac{e^{(\Gamma - i\Omega)t/2}}{D}[(1 + \kappa_A^2 e^{2i\omega t})\{(1 - \kappa_A\kappa)^2 e^{-\mu t} + (\kappa_A + \kappa)^2 e^{\mu t}\}$$
$$- 2(1 - e^{2i\omega t})\kappa_A(1 - \kappa_A\kappa)(\kappa_A + \kappa)\sinh\mu t], \qquad (33a)$$

$$J_{12} = \frac{-2e^{(\Gamma - i\Omega)t/2}}{D}[(1 + \kappa_A^2 e^{2i\omega t})(1 - \kappa_A\kappa)(\kappa_A + \kappa)\sinh\mu t$$
$$- \tfrac{1}{2}(1 - e^{2i\omega t})\kappa_A\{(1 - \kappa_A\kappa)^2 e^{\mu t} + (\kappa_A + \kappa)^2 e^{-\mu t}\}], \quad (33b)$$

$$J_{21} = \frac{-2e^{-(\Gamma - i\Omega)t/2}}{D}[(1 + \kappa_A^2 e^{-2i\omega t})(1 - \kappa_A\kappa)(\kappa_A + \kappa)\sinh\mu t$$
$$+ \tfrac{1}{2}(1 - e^{-2i\omega t})\kappa_A\{(1 - \kappa_A\kappa)^2 e_{\mu t} + (\kappa_A + \kappa)^2 e^{\mu t}\}], \quad (33c)$$

$$J_{22} = \frac{e^{-(\Gamma - i\Omega)t/2}}{D}[(1 + \kappa_A^2 e^{-2i\omega t})\{(1 - \kappa_A\kappa)^2 e^{\mu t} + (\kappa_A + \kappa)^2 e^{-\mu t}\}$$
$$+ 2(1 - e^{-2i\omega t})\kappa_A(1 - \kappa_A\kappa)(\kappa_A + \kappa)\sinh\mu t]. \qquad (33d)$$

The denominator is

$$D = (1 + \kappa_A^2)(1 + \kappa_A^2 + \kappa^2 + \kappa_A^2\kappa^2) \qquad (33e)$$

and the "precession frequency" is

$$\mu = \sqrt{\tfrac{1}{4}\{(\Gamma + i\Omega')(1 + \theta_A^2)^{1/2} - 2i\omega\}^2 - V^2}. \qquad (33f)$$

Although the preceding matrix elements are not especially aesthetic mathematically nor transparent physically, they nevertheless provide a closed-form analytical solution that is not only manageable (in comparison with RWA theory or numerical integration) but also highly accurate, as will be illustrated shortly. Correspondence between OFT and RWA solutions can be established by expanding the radical $(1 + \theta_A^2)^{1/2}$ in Eq. (33f) in a Taylor series in θ_A,

$$(1 + \theta_A^2)^{1/2} \sim 1 + \tfrac{1}{2}\theta_A^2 - \tfrac{1}{8}\theta_A^4 + \cdots.$$

If only the first term is kept—that is, if in effect one sets $\theta_A = 0$—then $\varepsilon^\pm = \varepsilon_R^\pm$ and $\kappa = \kappa_R$, $\kappa_A = 0$, whereupon the OFT solution reduces to the RWA solution given by Eqs. (23) and (24).

If the first two terms of the expansion are retained, however, the eigenvalues $\varepsilon^\pm$ take the same form as those of $\varepsilon_R^\pm$,

$$\varepsilon^\pm = -\tfrac{1}{4}(\gamma_1 + \gamma_2) \pm \sqrt{\tfrac{1}{4}\{(\Gamma - \delta\Gamma) - i(\Omega - \delta\Omega)\}^2 - V^2}, \quad (34)$$

except that there is now a frequency shift (generally termed an "ac Stark shift")

$$\delta\Omega = \frac{2V^2\Omega'}{\Gamma^2 + \Omega'^2} \quad (35)$$

and a rf field-dependent change in the decay rates

$$\delta\Gamma = \frac{2V^2\Gamma}{\Gamma^2 + \Omega'^2}. \quad (36)$$

At resonance the frequency shift becomes $\delta\Omega_{\omega=\omega_0} = 4V^2\omega_0/(\Gamma^2 + 4\omega_0^2)$, which in the case of stable levels (or levels of equal lifetime) reduces to the result $\delta\Omega_{\omega=\omega_0}^{\gamma_1=\gamma_2=0} = V^2/\omega_0$ obtained by Bloch and Siegert.

2.4 OCCUPATION PROBABILITIES

From Eqs. (23) and (32) it follows that the RWA and OFT state vectors of a two-level atom interacting with an oscillating field take the general forms:

(a) OFT solution

$$|\Psi_{OF}\rangle = e^{-(i\omega_1 + \frac{1}{2}\gamma_1)t}\left(J_{11}c_1^0 + J_{12}c_2^0\right)|1\rangle$$
$$+ e^{-(i\omega_2 + \frac{1}{2}\gamma_2)t}\left(J_{21}c_1^0 + J_{22}c_2^0\right)|2\rangle, \quad (37)$$

(b) RWA solution

$$|\Psi_{RW}\rangle = e^{-(i\omega_1 + \frac{1}{2}\gamma_1)t}\left(I_{11}c_1^0 + I_{12}c_2^0\right)|1\rangle$$
$$+ e^{-(i\omega_2 + \frac{1}{2}\gamma_2)t}\left(I_{21}c_1^0 + I_{22}c_2^0\right)|2\rangle. \quad (38)$$

Here t is the length of time in which the atom has been subjected to the applied rf electric field.

In this section we will examine the occupation probabilities $P_1(t) = |\langle 1|\Psi(t)\rangle|^2$ and $P_2(t) = |\langle 2|\Psi(t)\rangle|^2$ predicted by the OFT and RWA solutions as functions of frequency ω (actually $\omega/2\pi$), time t, and interaction strength V, and compare these results with those obtained by numerical integration of the Schrödinger equation. (A procedure for solving the Schrödinger equation numerically will be outlined later when the problem of the coupling of an arbitrary number of states by an oscillating field is taken up.) For simplicity, and as a matter of frequent experimental interest, we assume that state $|1\rangle$ is initially populated ($c_1^0 = 1$) and state $|2\rangle$ is initially unpopulated ($c_2^0 = 0$). Then, the occupation probabilities predicted by the two theoretical approaches are respectively

$$P_1^{(OF)}(t) = |J_{11}|^2 e^{-\gamma_1 t}, \qquad P_2^{(OF)}(t) = |J_{21}|^2 e^{-\gamma_2 t}, \qquad (39)$$

$$P_1^{(RW)}(t) = |I_{11}|^2 e^{-\gamma_1 t}, \qquad P_2^{(RW)}(t) = |I_{21}|^2 e^{-\gamma_2 t}. \qquad (40)$$

As a matter of terminology, P_2 in this case is ordinarily referred to as a transition probability since only the off-diagonal interaction matrix element J_{21} enters the expression. The distinction between occupation and transition probabilities loses its significance, however, when both states are initially populated, as is generally the case for collisional excitation of an atom.

As a concrete example of two states of substantially different lifetimes, let us identify $|1\rangle$ and $|2\rangle$ with the $4S_{1/2}$ and $4P_{1/2}$ states of hydrogen which, along with other hydrogenic S and P states, have been the subject of various resonance experiments to determine the Lamb shift and fine structure constant. The respective lifetimes are 232 ns for $4S_{1/2}$ and 12.4 ns for $4P_{1/2}$.

Figure 2.2 shows the occupation probability of the longer-lived state $|1\rangle$ calculated as a function of applied frequency by the three approaches (RWA, OFT, numerical). The exact lineshape follows a Lorentzian-like curve slightly shifted beyond the resonance frequency ω_0 (equal, in this example, to $2\pi \times 140$ MHz) by the counter-rotating component of the oscillatory field. Both the OFT and RWA yield nearly identical lineshapes (although the former includes the small frequency shift). For the long-lived state there are no major structural differences in the three lineshapes, a feature perhaps not unexpected.

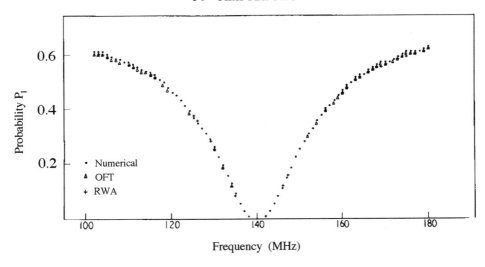

Figure 2.2 Occupation probability of a long-lived state according to OFT, RWA, and numerical solutions of the Schrödinger equation. Any plotting symbol not shown in the immediate vicinity of a dot is assumed coincident with the dot. Plotting parameters: $V/\omega_0 = 5/140$; $t = 80$ ns, $\gamma_1 = 4.35 \times 10^6$ s^{-1}, $\gamma_2 = 80.6 \times 10^6$ s^{-1}.

The outcome is quite different, however, in a comparison of probabilities of transition to the shorter-lived state $|2\rangle$ illustrated in figure 2.3. Numerical solutions, computed for two different values of the interaction time t, exhibit a high-frequency modulation of about 4 MHz and 6 MHz, respectively. This additional structure, while reproduced quite well by the OFT solution, is due solely to the counter-rotating component of the oscillating field and therefore absent from the RWA lineshape. Here, then, is an example of the effect of the counter-rotating component manifested throughout the *entire* lineshape and not merely as a small displacement of the peak. If $\Delta\omega$ represents the separation between modulation maxima, then from the figure we find that $(\Delta\omega)t = 1/2$.

To understand the foregoing results, let us look more closely at the probability of populating state $|2\rangle$ as prescribed by relation (33c). The functional form of J_{21} is fairly complicated, but in order to make qualitative statements it is permissible to retain only factors linear in κ_A since $\kappa_A \ll \kappa \ll 1$. If, in addition, we approximate the OFT functions μ, κ by the corresponding RWA functions ν, κ_R, then following the develop-

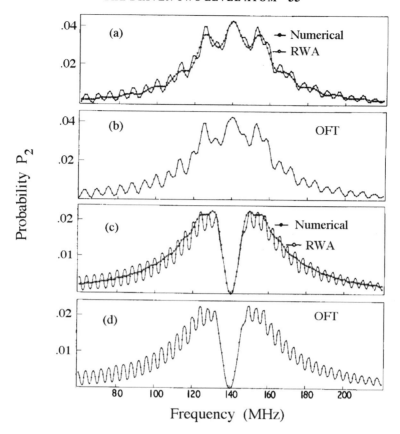

Figure 2.3 Occupation probability of a short-lived state as determined by OFT, RWA, and numerical integration for interaction times of 80 ns [(a), (b)] and 125 ns [(c), (d)]. All other parameters are the same as for figure 2.2.

ment of Section 2.2, the amplitude $\langle 2|\Psi_{OF}(t)\rangle$ can be written as

$$
\langle 2|\Psi_{OF}(t)\rangle \approx -2\exp\left\{-\left(\frac{\gamma_1 + \gamma_2}{2} + i\Omega\right)\frac{t}{2}\right\}
$$
$$
\times \left[\left(\frac{iV}{2\nu}\right)\sinh\nu t + \frac{1}{2}\left(1 - e^{-2i\omega t}\right)\kappa_A\cosh\nu t \right.
$$
$$
\left. - \left(\frac{\Gamma - i\Omega}{2\nu}\right)\sinh\nu t\right].
$$

$$(41)$$

With terms linear in κ_A again retained, the amplitude (41) leads to a transition probability

$$P_2^{(OF)}(t) \approx P_2^{(RF)}(t)$$
$$+ \frac{1}{\Omega'} \operatorname{Im}\{(1 - e^{-2i\omega t})[2\nu \coth \nu t - \Gamma + i\Omega]\}P_2^{(RF)}(t) \quad (42)$$

in which

$$P_2^{(RW)}(t) = 4\exp[\tfrac{1}{2}(\gamma_1 + \gamma_2)t]\left|\frac{iV}{2\nu}\sinh \nu t\right|^2. \quad (43)$$

The total lineshape is thus the sum of the RWA lineshape and a smaller modulated term inversely proportional to $\Omega' = \omega + \omega_0$. Since $e^{2i\omega t} = e^{2it(\omega + \pi/t)} = e^{2it(\omega + \Delta\omega)}$, we see that $(\Delta\omega/2\pi)t = \tfrac{1}{2}$ as observed from figure 2.3.

To examine the effect of level instability on a resonance lineshape, let us first consider the function $P_2^{(RW)}(t)$ in the case of stable states. Equation (24c) then leads to the familiar formula for "Rabi flips" in magnetic resonance:

$$P_2^{(RW)}(t) = \frac{V^2}{\tfrac{1}{4}\Omega^2 + V^2}\sin^2\left(\sqrt{\tfrac{1}{4}\Omega^2 + V^2}\,t\right), \quad (44a)$$

which, for small perturbation V, yields the result of first-order time-dependent perturbation theory,

$$P_2^{(RW)}(t) \approx (Vt)^2\frac{\sin^2(\tfrac{1}{2}\Omega t)}{(\tfrac{1}{2}\Omega t)^2}. \quad (44b)$$

Since the "sinc" function $(\sin x)/x$ goes smoothly to 1 in the limit $x \to 0$, there is no singularity at resonance. The resulting lineshape, qualitatively similar to curves (a) and (b) of figure 2.4, has a large central maximum which drops off steeply on both sides into oscillations with a frequency interval given by $(\Delta\omega)t = 1$. These will be designated as primary oscillations. The full width at half maximum varies roughly as $1/t$ and suggests that the lineshape can be narrowed in principle to a delta function by making the interaction time $t \to \infty$. The counter-rotating component of

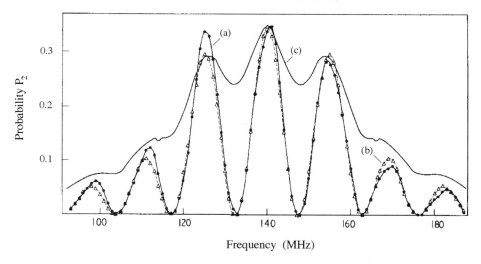

Figure 2.4 Effect of decay on the occupation probability of short-lived state. (a) Numerical solution for stable levels; (b) RWA solution for stable levels; (c) RWA solution for decaying levels (with parameters of figure 2.2).

the oscillatory field will cause small distortions and shifts in these primary oscillations, but the size of these effects is not great enough to alter substantially the narrow undulatory RWA lineshape.

For decaying states, the situation is different. Each state has a natural width given by the inverse lifetime. A resonance experiment coupling two such states will lead to a lineshape with an intrinsic width of at least $|\gamma_1 - \gamma_2|/2\pi$ for t greater than the shorter of the two lifetimes.[8] Such a lineshape cannot be further narrowed even in principle by letting the interaction time become arbitrarily large. Figure 2.4 shows a comparison of lineshapes for stable and decaying states. One sees that the broadening of the line due to the natural decay of the states greatly attenuates the primary oscillations. Secondary oscillations attributable to the modulation term in Eq. (42) then become a significant feature.

In addition to the variation in occupation probabilities with applied frequency, it is instructive to consider the evolution of the atom in time as well. Figure 2.5 shows the temporal dependence of $P_2(t)$ at resonance ($\omega = \omega_0$). The numerical solution shows that there are again low-frequency, large-amplitude primary oscillations and high-frequency, small-amplitude secondary oscillations. If the interaction V is expressed in units of angular frequency, the period of the primary oscillations is

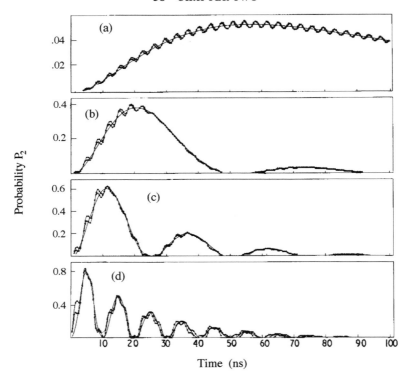

Figure 2.5 Variation with time of the occupation probability of a short-lived state as determined by RWA (smooth line), OFT and numerical integration (undulatory lines—virtually indistinguishable at scale shown). (a) $V = 2$ MHz; (b) $V = 10$ MHz; (c) $V = 20$ MHz; (d) $V = 50$ MHz. Other parameters are the same as for figure 2.2. At the scale shown, the OFT and exact solutions are almost indistinguishable.

given by $V(\Delta t) = \pi$. Each succeeding local maximum is smaller than the previous one by the decay factor $\exp[-\frac{1}{2}(\gamma_1 + \gamma_2)t]$. For V smaller than a critical value there are no primary oscillations, but only one extremely broad peak. The frequency of the secondary oscillations is $2\omega_0$ and is insensitive to V. This behavior is reproduced by the OFT solution, whereas the RWA solution shows only the primary oscillations.

We can interpret the preceding features by resorting again to Eqs. (42)–(44). In the case of stable states at resonance, $P_2(t)$ reduces simply to $\sin^2 Vt$, and therefore oscillates harmonically between 0 and 1 with a period given by $V(\Delta t)\pi$. With resonance established between decaying states, however, there are effectively two different behaviors depending on the size of the interaction compared with the level decay rates. If $V^2 > \frac{1}{4}\Gamma^2$, then $P_2^{(RW)}(t)$ is an exponentially decaying oscillatory function proportional to $\sin^2(\sqrt{V^2 - \frac{1}{4}\Gamma^2}t)$ with a recurrence time Δt that depends on the decay rate difference

$$V(\Delta t) = \frac{\pi}{\sqrt{1 - \dfrac{\Gamma^2}{V^2}}}. \tag{45}$$

The shape of the transient is qualitatively the same as for stable states except for a reduction of each primary maximum by the factor $\exp[-\frac{1}{2}(\gamma_1 + \gamma_2)t]$. If, however, $V^2 < \frac{1}{4}\Gamma^2$, the results are quite different. The transition probability, now proportional to the square of a hyperbolic sine, is not periodic, but has a single maximum whose location can be deduced from solving the equation $dP_2^{(RW)}(t)/dt = 0$, which leads to the transcendental relation

$$\tanh\left(\sqrt{\tfrac{1}{4}\Gamma^2 - V^2}t\right) = \frac{4\sqrt{\tfrac{1}{4}\Gamma^2 - V^2}t}{\gamma_2 + \gamma_2}. \tag{46}$$

In both cases there is a $2\omega_0$ modulation arising from the counter-rotating field.

Finally, we examine the variation in P_1 and P_2 with the remaining experimentally adjustable parameter, the interaction strength V. In figure 2.6 are displayed the results of numerical integration for the condition of resonance (and a fixed interaction time of 50 ns). The OFT and RWA solutions yield virtually identical results in these cases and are not shown separately.

For the long-lived state, the occupation probability is $P_1(t) = e^{-\gamma_1 t}$ at $V = 0$ as determined by the initial conditions; it rapidly decreases with increasing V and becomes oscillatory with a recurrence relation $(\Delta V)t = \pi$. For the short-lived state, $P_2(t)$ rises from 0 at $V = 0$, reaches a maximum, and then breaks into similar oscillations. Secondary oscillations do not appear in these variations because both the interaction time and applied frequency are constant.

One noteworthy characteristic of the transitions driven between two stable states is that the total probability of finding an atom in one state or the other is always 100%. The conservation of probability is a general

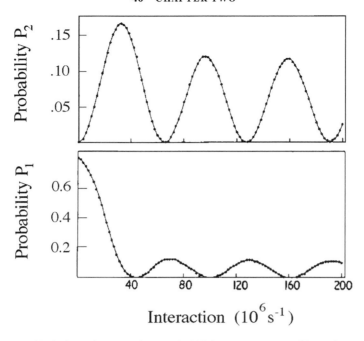

Figure 2.6 Variation of occupation probabilities at resonance ($\Omega = 0$) with coupling strength. Interaction time = 50 ns; other parameters are the same as for figure 2.2.

property of the Schrödinger equation provided the Hamiltonian is Hermitian. One sees it explicitly in the RWA occupation and transition probabilities at resonance in which $P_1(t) = \cos^2(Vt)$ and $P_2(t) = \sin^2(Vt)$. Experimentally, the significance of the preceding two expressions (which sum identically to unity) is that the coupling strength V at which state $|1\rangle$ is maximally quenched is the same strength at which state $|2\rangle$ is maximally pumped.

This seemingly obvious reciprocity no longer holds for coupled unstable states. The Hamiltonian is not Hermitian, for there was inserted "by hand" at the outset the anti-Hermitian term leading to state decay. Moreover, a state decays (by spontaneous emission) to others that lie *outside* the manifold of states comprising the coupled system of interest. Total probability, therefore, is not conserved, and one consequence of this is that maximal pumping and quenching can occur at different values of the coupling parameter V. For example, based on the RWA solutions (which are suitably accurate in the present case), one can show that the extrema of the occupation probabilities are determined from the following transcendental equations:

(a) $V^2 > \frac{1}{4}\Gamma^2$

$$P_{1\,\text{min}}^{(\text{RW})} \Rightarrow \tan(Yt) = \frac{2Y}{\Gamma},$$
$$P_{2\,\text{max}}^{(\text{RW})} \Rightarrow \tan(Yt) = (Yt)\left(1 + \frac{4Y^2}{\Gamma^2}\right)$$

(47a)

with $Y = \sqrt{V^2 - \frac{1}{4}\Gamma^2}$, and

(b) $\frac{1}{4}\Gamma^2 > V^2$

$$P_{1\,\text{min}}^{(\text{RW})} \Rightarrow \tanh(Yt) = \frac{2Y}{\Gamma},$$
$$P_{2\,\text{max}}^{(\text{RW})} \Rightarrow \tanh(Yt) = (Yt)\left(1 - \frac{4Y^2}{\Gamma^2}\right)$$

(47b)

with $Y = \sqrt{\frac{1}{4}\Gamma^2 - V^2}$. Only in the case of two states of equal lifetime ($\Gamma = 0$) do $P_1(t)$ and $P_2(t)$ reach extremal values for the same applied field strength.

It is useful to summarize the principal lessons of this chapter as follows. In atomic and molecular beam experiments limited to stable or metastable states the simple rotating-wave approximation often suffices. However, rf spectroscopy of transient states, which fast-beam techniques make possible, usually requires a more thorough understanding of the effects of an oscillating-field. The theoretical approach, termed the oscillating field theory, described in this chapter takes account of both the resonant and antiresonant frequency components of the rf field and leads to an approximate analytical solution to the Schrödinger equation in close agreement with the results of exact numerical integration.

Depending on the magnitude of the difference in state decay rates compared with the interaction matrix element, electric resonance lineshapes of transient states may be qualitatively different from those of stable states and manifest structural features (attributable to the antiresonant field) not reproduced by the rotating-wave approximation. One such feature of particular importance to precision measurements of atomic level separations is the displacement of the apparent resonance frequency from the Bohr frequency, given by Eq. (35). This result, which generalizes the Bloch-Siegert shift for stable states, has been subsequently confirmed.[9]

NOTES

1. F. Bloch and A. Siegert, "Magnetic Resonance for Nonrotating Fields," *Phys. Rev.* **57** (1940):522

2. M. P. Silverman and F. M. Pipkin, "Interaction of a Decaying Atom with a Linearly Polarized Oscillating Field," *J. Phys. B: Atom. Molec. Phys.* **5** (1972):704.

3. A Hermitian operator H is equal to its adjoint $H^\dagger$. The significance of such operators is that their eigenvalue spectrum is real valued. An anti-Hermitian operator A is equal to $-A^\dagger$ and has an imaginary-valued eigenvalue spectrum.

4. Technically, Eq. (4) is known as the Schrödinger *form* of the equation of motion, in contrast to the Heisenberg *form* which expresses the time evolution of operators. The actual Schrödinger equation is a differential equation for the wavefunction, obtained by projecting the state vector onto an appropriate set of basis states.

5. M. P. Silverman, *More than One Mystery: Explorations in Quantum Interference* (Springer, New York, 1995), chap. 5.

6. M. P. Silverman, *Waves and Grains: Reflections on Light and Learning* (Princeton University Press, Princeton, NJ, 1998).

7. The value of ε_{ijk} is $+1$ if (i, j, k) form an even permutation of $(1, 2, 3)$ (as in ε_{312}) and -1 if the permutation is odd (as in ε_{132}); the symbol takes the value 0 if any two indices are repeated. The symbol δ_{ij} takes the value $+1$ whenever the two indices are equal, and zero otherwise.

8. It has sometimes been mistakenly believed that the resonance linewidth should depend on the sum of the decay rates of the states, rather than on the difference, for this is known to be the case in optical spectroscopy where one observes spontaneous emission from uncoupled states. Under circumstances in which radiative decay accompanies transitions between states driven by an external oscillating field, there is no reason to expect the same result.

9. D. A. Andrews and G. Newton, "Observation of the Bloch-Siegert shifts in the $2\,^2S_{1/2}$–$^2P_{1/2}$ microwave resonance in atomic hydrogen," *J. Phys. B: Atom. Molec. Phys.* **8** (1975):1415.

CHAPTER 3

The Driven Multilevel Atom

3.1 STATISTICAL UNCERTAINTIES AND THE DENSITY MATRIX

A pure quantum mechanical state, often referred to as a state of maximal information, is one for which there is a definite wavefunction, unique up to a unitary transformation.[1] The two-level theory of the preceding chapter treated such a pure system; initial conditions of both the atom and the applied field were presumed known; there were no additional uncertainties apart from the elements of uncertainty and probability intrinsic to quantum mechanics itself.

In many experiments, however, one must deal with states of less than maximal information[2]—mixed states that can be described by the incoherent superposition of pure states. For these situations the statistical or density operator (familiarly termed the density matrix, even though matrices constitute only one explicit representation of an operator) provides a convenient formalism with which to handle both quantum and classical uncertainties.

For the experimental circumstances addressed in this book, involving the excitation of atoms subsequently passing through time-dependent fields, incompleteness of information can arise in at least two ways. First, there is an uncertainty in the phase of the quantum mechanical amplitudes characterizing the state of each atom at the instant of its preparation. Second, there is a phase uncertainty due to the entry of different atoms into the rf field at different points in the oscillation cycle. It will be useful to make a distinction between those distributed variables associated with the creation process and those introduced by the process of investigation. To do this, let us represent by σ the statistical operator obtained from an ensemble average of pure states over initial quantum mechanical phases ϕ; $\langle \ \rangle_\phi$ signifies this averaging operation. The symbol ρ will represent the statistical operator obtained from an ensemble average of σ over all distributed variables δ arising from the measurement process; $\langle \ \rangle_\delta$ signifies this averaging procedure. In contrast to experiments with thermal beams, it is unnecessary here to average over a distribution of atomic velocities since the accelerator produces a highly monoenergetic beam of charged particles whose conversion to fast neutral atoms creates little energy dispersion.

Consider first the case in which the target is a very thin foil. (Carbon foils, for example, have been widely used for this purpose.) At the instant $t = 0$ a proton interacts with a target atom resulting in the capture of an outer electron. Later in the book a specific model of the capture process based on a Coulombic spin-independent interaction will be delineated, but for the present let us simply represent the wavefunction of the resulting atom as a superposition of certain basis states:

$$|\Psi_0\rangle = \sum_\alpha Q_\alpha|\alpha\rangle, \tag{1}$$

where α signifies a set of quantum numbers uniquely defining each state vector. The dependence of the magnitude q_α and phase ϕ_α of each amplitude Q_α,

$$Q_\alpha = q_\alpha e^{i\phi_\alpha}, \tag{2}$$

on the quantum numbers comprising the state label, will vary according to the particular model of atom production.

In principle one is free at the outset to select any complete basis set to describe a quantum mechanical system as, for example, the set of uncoupled orbital, electron spin, and nuclear spin states $|nLM_L\rangle|SM_S\rangle|IM_I\rangle \equiv |nLSIM_LM_SM_I\rangle$, or the set of coupled angular momentum states $|nLSJIFM_F\rangle$ [$n^{2S+1}L_{2J+1}(FM_F)$ in spectroscopic notation] constituting the eigenstates of the field-free Hamiltonian.[3] When, however, assumptions are made regarding the correlations between phases of the amplitudes Q_α, the selected basis is given a certain preference, and predictions of measurable quantities such as the polarization and angular distribution of emitted radiation will reflect those assumptions. What basis provides the most suitable description of a physical process must be answered experimentally. As already intimated above, this issue will be examined in more detail in a later section devoted to electron capture. There, both the uncoupled and coupled bases will play a role, for an electron is captured into states of the former, but undergoes induced transitions and radiative decay from states of the latter.

The selection of a basis also entails defining an axis of quantization—that is, the axis with respect to which angular momentum is quantized. This choice is arbitrary in the sense that the prediction of physical observables cannot depend upon it. However, some choices are clearly more convenient than others. In the absence of external fields, it is usual to take the beam direction to be the quantization axis since it is the only symmetry axis available. Treatment of the interaction of atomic states with an external field, however, is usually markedly simplified by choosing the quantization axis along the field direction. For reasons of experimental convenience the radiofrequency field is often oriented perpendicular to the beam direction, and thus there is a conflict as to the most appropriate choice. In this book quantum states are usually defined initially with respect to the beam and, when necessary, subsequently transformed to a basis defined with respect to an applied field.

From Eq. (1), the statistical operator for an ensemble of similarly prepared atoms takes the form

$$\sigma^0 = \langle |\Psi_0\rangle\langle\Psi_0|\rangle_\phi = \sum_{\alpha,\beta} |\alpha\rangle\sigma^0_{\alpha\beta}\langle\beta| \tag{3}$$

where

$$\sigma^0_{\alpha\beta} = \langle Q_\alpha Q^*_\beta\rangle_\phi. \tag{4a}$$

The diagonal elements

$$\sigma^0_{\alpha\alpha} = \langle |q_\alpha|^2\rangle_\phi \tag{4b}$$

represent relative state populations, whereas off-diagonal elements $(\alpha \neq \beta)$

$$\sigma^0_{\alpha\beta} = \langle q_\alpha q_\beta e^{i(\phi_\alpha - \phi_\beta)}\rangle, \tag{4c}$$

indicative of "coherence terms," survive averaging only if there exist specific nonrandom relationships among the amplitude phases. It is the existence of such coherence terms that leads, for example, to quantum interference effects ("quantum beats") in the spontaneous emission of atoms and molecules.

3.2 TIME EVOLUTION OF THE DENSITY MATRIX

The equation of motion of the statistical operator σ (or ρ) in the Schrödinger representation is[4]

$$\frac{d\sigma(t)}{dt} = i[\sigma(t), \mathcal{H}_a] - i\{\sigma(t), \mathcal{H}_D\} = i[\sigma(t), \mathcal{H}_a] - \tfrac{1}{2}\{\sigma(t), \Gamma_D\}, \quad (5)$$

where the total Hamiltonian $\mathcal{H}$ is the sum of a Hermitian term ($\mathcal{H}_a = \mathcal{H}_a^\dagger$) generating the atomic energy spectrum and transitions between energy states and an anti-Hermitian Hamiltonian ($\mathcal{H}_D = -i\tfrac{1}{2}\Gamma_D$, with $\Gamma_D = \Gamma_D^\dagger$) responsible for level decay. In Eq. (5) square brackets signify operator commutation $[A, B] = AB - BA$, and curly brackets signify anticommutation $\{A, B\} = AB + BA$. If the Hamiltonian is independent of time, then Eq. (5) can be integrated immediately to yield a formal solution:

$$\sigma(t) = e^{-(i\mathcal{H}_a + \frac{1}{2}\Gamma_D)(t-t_0)} \sigma(t_0) e^{(i\mathcal{H}_a - \frac{1}{2}\Gamma_D)(t-t_0)}. \quad (6)$$

If $\mathcal{H}_a$ is the field-free Hamiltonian $\mathcal{H}_0$ of the preceding chapter, the density matrix elements, expressed in a basis of the eigenstates of $\mathcal{H}_0$, can be written as

$$\sigma_{\alpha\beta}(t) = e^{-(i\omega_{\alpha\beta} + \frac{1}{2}\gamma_{\alpha\beta})(t-t_0)} \sigma(t_0), \quad (7)$$

in which

$$\omega_{\alpha\beta} = \omega_\alpha - \omega_\beta, \quad (8)$$

$$\gamma_{\alpha\beta} = \gamma_\alpha + \gamma_\beta. \quad (9)$$

[Note the sign difference between Eqs. (8) and (9).] The diagonal elements of Eq. (7),

$$\sigma_{\alpha\alpha}(t) = e^{-\gamma_\alpha(t-t_0)} \sigma_{\alpha\alpha}(t_0), \quad (10)$$

express the simple exponential decay of the state populations.

Let us consider next the effect of an extended target, that is, one in which the atom production occurs over a finite length ℓ as shown in figure 3.1. A proton entering from the left is converted to a hydrogen

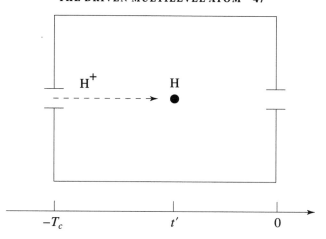

Figure 3.1 Schematic diagram of atom production in an extended target. Electron capture by H^+ to form neutral H occurs throughout the time interval $(-T_c, 0)$.

atom within the target at time t' where the time origin $t = 0$ is taken to be the point of emergence of the atom from the collision chamber. The maximum time an atom can spend within the chamber is $T_c = \ell/v$ where v is the velocity of the atom. A density matrix element at time $t = 0$ for one atom created at time t' earlier is therefore

$$\sigma_{\alpha\beta}(0, t') = e^{(i\omega_{\alpha\beta} + \frac{1}{2}\gamma_{\alpha\beta})t'} \sigma_{\alpha\beta}(t'). \tag{11}$$

To obtain the element for an ensemble of atoms we must average Eq. (11) over the range $-T_c \le t' \le 0$,

$$\begin{aligned}
\langle \sigma_{\alpha\beta}(0, t') \rangle_{t'} &= \frac{1}{T_c} \int_{-T_c}^{0} \sigma_{\alpha\beta}(0, t') \, dt' \\
&= \frac{1}{T_c} \int_{-T_c}^{0} e^{(i\omega_{\alpha\beta} + \frac{1}{2}\gamma_{\alpha\beta})t'} \sigma_{\alpha\beta}(t') \, dt'. \tag{12}
\end{aligned}$$

As the space within the target is assumed homogeneous, the initial density matrix elements must be independent of the point of creation, and one can therefore set $\sigma_{\alpha\beta}(t') = \sigma_{\alpha\beta}^0$; Eq. (12) then reduces to

$$\langle \sigma_{\alpha\beta}(0, t') \rangle_{t'} = \Phi\{(i\omega_{\alpha\beta} + \tfrac{1}{2}\gamma_{\alpha\beta})T_c\}\sigma_{\alpha\beta}^0, \tag{13a}$$

where we have defined the function

$$\Phi\{x\} \equiv \frac{1 - e^{-x}}{x}. \tag{13b}$$

In the limit that the target extension vanishes, as in the case of a foil target, $\Phi\{x\}$ approaches unity. In the following discussion it will be understood that the averaging procedure expressed by Eq. (12) has always been performed; we will therefore dispense with the notation $\langle \ \rangle_{t'}$ and simply write

$$\sigma_{\alpha\beta}(0) = \Phi\{(i\omega_{\alpha\beta} + \tfrac{1}{2}\gamma_{\alpha\beta})T_c\}\sigma^0_{\alpha\beta}. \tag{13c}$$

3.3 GENERALIZED RESONANT FIELD THEORY

In contrast to chapter 2, which focused on a single atom in a pure quantum state, the problem of a beam of atoms interacting with an oscillating field requires that we take account of distributed or uncertain variables arising from generation and measurement. For this purpose the density matrix formalism was introduced in the previous section. Although there are situations (as, for example, the excitation of atoms by wide-band laser illumination[5]) for which the density matrix elements are most readily obtained directly from the equation of motion (5), in the present case it is actually more convenient to begin with the Schrödinger equation for quantum amplitudes.

Let us represent the state of a single atom at time t as a superposition of N field-free eigenstates weighted by time-dependent coefficients

$$|\Psi(t)\rangle = \sum_{\mu=1}^{N} c_\mu(t)|\mu\rangle \tag{14}$$

to be determined from the Schrödinger equation

$$\mathcal{H}|\Psi(t)\rangle = i\frac{\partial|\Psi\rangle}{\partial t}. \tag{15}$$

The total Hamiltonian $\mathcal{H}$,

$$\mathcal{H} = \mathcal{H}_0 + \mathcal{H}_D - \boldsymbol{\mu}_E \cdot \mathbf{E}_0 \cos(\omega t + \delta), \tag{16}$$

comprises terms for the field-free energy, radiative decay, and electric dipole interaction with an incident rf field. Equation (16) differs from Eq. (2.1) of the previous chapter only by inclusion of a phase δ which determines the magnitude of the field initially encountered by each entering atom, a distributed variable since atoms arrive in the rf chamber at different times throughout the oscillation cycle.

Substitution of Eq. (14) into Eq. (15) generates the following set of coupled differential equations (for $\mu = 1$ to N):

$$
\begin{aligned}
\frac{dc_\mu(t)}{dt} = &-(i\omega_\mu + \tfrac{1}{2}\gamma_\mu)c_\mu(t) \\
&- i\sum_{\nu=1}^{N} c_\nu(t)V_{\mu\nu}\big(e^{i(\omega t+\delta)} + e^{-i(\omega t+\delta)}\big)
\end{aligned}
\tag{17}
$$

with interaction matrix elements

$$
V_{\mu\nu} = \frac{\langle \mu| - \boldsymbol{\mu}_E \cdot \mathbf{E}_0|\nu\rangle}{2\hbar}.
\tag{18}
$$

As a generalization of Eq. (2.9) defining the interaction representation for a two-level system, the transformation

$$
c_\mu(t) = a_\mu(t)e^{-i\omega_\mu t}
\tag{19}
$$

removes the high-frequency part of the first term of Eq. (17), leading to

$$
\frac{da_\mu(t)}{dt} = -\tfrac{1}{2}\gamma_\mu a_\mu(t) - i\sum_{\nu=1}^{N} a_\nu(t)V_{\mu\nu}\big(e^{i(\Omega_{\mu\nu}t+\delta)} + e^{-i(\Omega'_{\mu\nu}t+\delta)}\big).
\tag{20}
$$

The frequencies

$$
\Omega_{\mu\nu} = \omega - \omega_{\mu\nu},
\tag{21a}
$$

$$
\Omega'_{\mu\nu} = \omega + \omega_{\mu\nu}
\tag{21b}
$$

respectively characterize the resonant and antiresonant terms when $\omega_{\mu\nu} > 0$, and the reverse when $\omega_{\mu\nu} < 0$. It will be assumed throughout this discussion that the levels are so ordered that $\omega_1 > \omega_2 > \cdots > \omega_N$, in which case $\Omega_{\mu\nu}$ with $\mu < \nu$ is always associated with the resonant component.

For a particular value of the applied frequency, only those states with an energy interval represented by $\omega \sim \omega_{\mu\nu}$ will be effectively coupled by the rf field provided $V_{\mu\nu}$ does not vanish. If M of the N levels are coupled by the rf field, then Eq. (20) reduces to a set of M coupled equations and $N - M$ uncoupled equations whose solutions are immediately given by

$$c_\mu(t) = e^{-(i\omega_\mu + \frac{1}{2}\gamma_\mu)(t-t_0)} c_\mu(t_0). \tag{22}$$

The number of states coupled by the oscillating field can be minimized by adopting the direction of the field $\mathbf{E}_0$ as the axis of quantization, and thus allowing only $\Delta M_F = 0$ transitions (where F is the total angular momentum quantum number). One can then express the density matrix in terms of a basis quantized along any other direction by means of a rotational transformation, as follows.

Let $\sigma^{(z)}(t)$ be the statistical operator of the beam at time t referred to a basis quantized along the beam (z) axis, and $\sigma^{(e)}(t)$ be defined with respect to the polarization (e) axis of the applied field. The two are related by

$$\sigma^{(z)} = D(z \to e)\sigma^{(e)}D^{-1}(z \to e), \tag{23}$$

where $D(z \to e)$ is the rotation operator whose argument signifies the set of Euler angles, usually designated (α, β, γ), for rotating the z axis into the e axis. The angles for this specific rotation are $(\phi, \theta, 0)$ in which θ and ϕ are, respectively, the polar and azimuthal angles of e with respect to the right hand coordinate system (x, y, z).

In the case of a time-independent Hamiltonian, $\sigma^{(e)}(t)$ is generated from the statistical operator at an earlier time t_0 by a time-evolution operator $U(t, t_0)$ through a transformation of the form

$$\sigma^{(e)}(t) = U(t, t_0)\sigma^{(e)}(t_0)U^\dagger(t, t_0). \tag{24}$$

[See Eq. (6).] Utilizing the inverse of Eq. (23) to obtain $\sigma^{(z)}(t_0)$, we can express $\sigma^{(z)}(t)$ explicitly as follows:

$$\sigma^{(z)}(t) = \{D(z \to e)U(t, t_0)D^{-1}(z \to e)\}\sigma^{(z)}(t_0)$$
$$\times \{D(z \to e)U^\dagger(t, t_0)D^{-1}(z \to e)\}. \tag{25}$$

A common experimental configuration in the rf spectroscopy of atomic beams is to have the field axis perpendicular to the beam axis, in which case $\theta = \pi/2$, $\phi = 0$.

As discussed in the previous chapter, no exact closed-form solution to Eq. (20) is known even in the case of a two-state system—and the inclusion of additional states does not make the problem more tractable. However, the ensemble averaging effected by passage of a beam of atoms through the oscillating field (in contrast, for example, to performing a resonance experiment on a single trapped atom or ion) smooths the lineshapes so that the secondary modulation of the antiresonant component is less prominent. Moreover, an atom entering or leaving the rf field does not encounter the field with the abruptness of a step function, but passes through a short transition zone; this, too, tends to average away effects of the counter-rotating field. The set of coupled equations (20) can be solved, therefore, by a generalization of the rotating-wave approximation employed in chapter 2. Since no field is actually rotating, it is preferable to designate this approach the "generalized resonant field" (GRF) theory.

The GRF solution entails discarding terms in $\Omega'_{\mu\nu}$ since these terms do not pass through an extremum as $\omega \rightarrow \omega_{\mu\nu}$ and hence do not lead to resonance effects, but rather to slight lineshape distortions and frequency shifts. Equation (20) then reduces to

$$\frac{da_\mu(t)}{dt} = -\tfrac{1}{2}\gamma_\mu a_\mu(t) - i\sum_{\nu=1}^{N} a_\nu(t)V_{[\mu\nu]}e^{[-]i(\Omega_{[\mu\nu]}t+\delta)}, \qquad (26)$$

where the bracketed subscripts signify that indices are always to be written in increasing numerical order (e.g., V_{13}, Ω_{13}) and the bracketed sign $[-]$ is negative for $\mu < \nu$ and positive for $\mu > \nu$.

The key to solving the set of equations (26) is to remove all time-dependent phase factors. In the case of the two-level system treated previously, this was accomplished by transforming the equation of motion into a frame "rotating" with the angular frequency Ω_{12}. With the addition of other levels, there is now no such obvious frame in which to transform Eq. (26). Let us assume, however, that the procedure is still possible and perform the following unitary transformation:

$$a_\mu(t) = e^{-i\lambda_\mu t}b_\mu(t), \qquad (27)$$

in which the (time-independent) elements λ_μ of a diagonal matrix Λ are to be determined. The transformed equation then takes the form

$$\frac{db\mu(t)}{dt} = \sum_{\nu=1}^{N}\left[-(\tfrac{1}{2}\gamma_\nu - i\lambda_\nu)\delta_{\mu\nu}\right.$$
$$\left. - i\left(V_{[\mu\nu]}e^{[-]i\delta}\right)e^{[-]i(\Omega_{[\mu\nu]}-\lambda_\mu+\lambda_\nu)t}\right]b_\nu(t) \quad (28a)$$

(where $\delta_{\mu\nu}$ is the Kronecker delta symbol), or in matrix notation,

$$\frac{db}{dt} = \Pi b, \quad (28b)$$

where b is a column vector of b-amplitudes, and the elements of the Hamiltonian Π are defined by the bracketed expression in Eq. (28a). If elements of Λ can be found such that

$$\lambda_\mu - \lambda_\nu = \Omega_{[\mu\nu]} \quad (29)$$

for all applicable resonant terms, then Π becomes a time-independent matrix, and the equation of motion is readily integrable.

The outcome of the integration and subsequent transformations back to the original Schödinger equation is the complete state vector

$$\Psi(t) = e^{-i(H_0+\Lambda)t}Ce^{\varepsilon(t-t_0)}C^{-1}e^{i(H_0+\Lambda)t}\Psi(t_0), \quad (30)$$

where $(H_0)_{\mu\nu} = \omega_\mu\delta_{\mu\nu}$ is the diagonal matrix of field-free energy eigenvalues (expressed as angular frequencies), and C is a matrix that brings Π to the diagonal form ε,

$$C^{-1}\Pi C = \varepsilon \quad (31)$$

[see Eq. (2.18)]. The elements of C and ε are obtained by standard techniques of linear algebra as employed in the previous chapter.

Can one always find a diagonal transformation matrix Λ whose elements satisfy Eq. (29)? In general, the answer is no. For a set of N levels there are N elements λ_μ, but the number of independent frequencies $\Omega_{[\mu\nu]}$ is $\tfrac{1}{2}N(N-1)$. For $N = 2$, there is but one frequency Ω_{12}, and Eq. (29) can be satisfied in any number of ways, for example, by setting $\lambda_1 = \Omega_{12}$ and $\lambda_2 = 0$. For $N = 3$, there are three eigenvalues

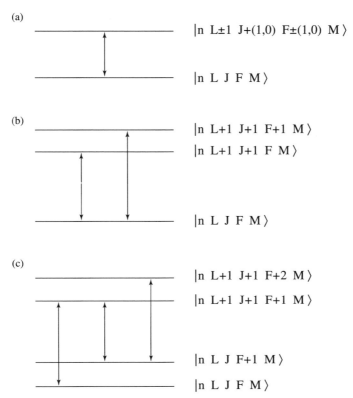

Figure 3.2 Allowable two-, three-, and four-level electric dipole transitions between nondegenerate states in hydrogen with the axis of quantization parallel to the applied field.

of Λ and three independent frequencies Ω_{12}, Ω_{13}, Ω_{23}—but the full set of three equations is not internally consistent. For $N = 4$ and higher, the number of frequencies $\Omega_{[\mu\nu]}$ to be removed is greater than the number of elements λ_μ. The procedure works, however, in the case of hydrogenic atoms (with the quantization axis chosen parallel to the rf electric field) because the number of allowable electric dipole transitions between nondegenerate states results in a smaller set of frequencies $\Omega_{[\mu\nu]}$ than available eigenvalues λ_μ.

As illustrated in figure 3.2, the rf electric field can simultaneously couple two, three, or four field-free hydrogen states of the same electronic manifold giving rise, respectively, to one-, two-, or three-state resonant transitions. Angular momentum selection rules lead to the vanishing of matrix elements $V_{[\mu\nu]}$—and hence the associated frequencies $\Omega_{[\mu\nu]}$—for all transitions that violate $\Delta F = 0,1$ and $\Delta M_F = 0$; transitions between $F = 0$ states are also forbidden, as an obvious violation of angular momentum conservation. We next examine the solutions and density matrix elements for these three kinds of couplings.

3.4 TWO-STATE TRANSITIONS

Setting $\lambda_1 = \frac{1}{2}\Omega_{12}$ and $\lambda_2 = -\frac{1}{2}\Omega_{12}$ as an alternative to the choice in the preceding section also removes the time dependence from the matrix Π, which then reduces to the RWA Hamiltonian of chapter 2:

$$\Pi^{(2)} = \begin{pmatrix} -\frac{1}{2}(\gamma_1 - i\Omega_{12}) & -iV_{12}e^{-i\delta} \\ -iV_{12}e^{i\delta} & -\frac{1}{2}(\gamma_2 + i\Omega_{12}) \end{pmatrix}, \qquad (32)$$

but with inclusion of the distributed phase δ. The solution, Eq. (30), of the two-state Schrödinger equation takes the form of Eq. (2.23):

$$\Psi(t) = \begin{pmatrix} e^{-(i\omega_1 + \frac{1}{2}\gamma_1)T} & 0 \\ 0 & e^{-(i\omega_2 + \frac{1}{2}\gamma_2)T} \end{pmatrix} \begin{pmatrix} I_{11} & I_{12} \\ I_{21} & I_{22} \end{pmatrix} \Psi(t_0), \qquad (33)$$

in which the set of interaction matrix elements $I_{\mu\nu}$.

$$I_{11} = e^{\frac{1}{2}(\Gamma - i\Omega_{12})T}\left[\cosh \nu T - \left(\frac{\Gamma - i\Omega_{12}}{2\nu}\right)\sinh \nu T\right], \qquad (34a)$$

$$I_{12} = -ie^{\frac{1}{2}(\Gamma - i\Omega_{12})T - i\omega t_0 - i\delta}\left(\frac{V_{12}}{\nu}\right)\sinh \nu t, \qquad (34b)$$

$$I_{21} = -ie^{-\frac{1}{2}(\Gamma - i\Omega_{12})T + i\omega t_0 + i\delta}\left(\frac{V_{12}}{\nu}\right)\sinh \nu t, \qquad (34c)$$

$$I_{22} = e^{-\frac{1}{2}(\Gamma - i\Omega_{12})T}\left[\cosh \nu T + \left(\frac{\Gamma - i\Omega_{12}}{2\nu}\right)\sinh \nu T\right], \qquad (34d)$$

denote explicitly the time interval t_0 between exiting the target and entering the rf field, and the duration T of passage through the field;

$t = T + t_0$ is the total time interval. As pointed out in chapter 2, the interaction elements $I_{\mu\nu}$ depend on the difference in decay rates,

$$\Gamma = \tfrac{1}{2}(\gamma_1 - \gamma_2). \tag{34e}$$

The expression

$$\nu = \sqrt{\tfrac{1}{4}(\Gamma - i\Omega_{12})^2 - V_{12}^2} \tag{34f}$$

is one-half the difference of the eigenvalues of $\Pi^{(2)}$ and, in the case of stable states ($\Gamma = 0$), corresponds physically (to within a factor $i = \sqrt{-1}$) to the precession frequency of a classical dipole in the vector model of a two-state system.[6]

The density matrix of the system is obtained from Eqs. (33) and (34a–d) by performing the ensemble average

$$\rho(t) = \langle \Psi(t)\Psi^{\dagger}(t) \rangle_{\phi,\,\delta}. \tag{35}$$

This leads to the matrix

$$\rho(t) = \begin{pmatrix} e^{-\gamma_1 T}\{|I_{11}|^2\sigma_{11}(t_0) + |I_{12}|^2\sigma_{22}(t_0)\} & e^{-(\frac{1}{2}\gamma_{12}+i\omega_{12})T}\{I_{11}I_{22}^*\sigma_{12}(t_0)\} \\ e^{-(\frac{1}{2}\gamma_{12}-i\omega_{12})T}\{I_{11}^*I_{22}\sigma_{21}(t_0)\} & e^{-\gamma_2 T}\{|I_{21}|^2\sigma_{11}(t_0) + |I_{22}|^2\sigma_{22}(t_0)\} \end{pmatrix} \tag{36}$$

in which the elements

$$\sigma_{\mu\mu}(t_0) = e^{-\gamma_\mu t_0}\Phi\{\gamma_\mu T_c\}\sigma_{\mu\mu}^0 \qquad (\mu = 1, 2), \tag{37a}$$

$$\sigma_{12}(t_0) = \sigma_{21}^*(t_0) = e^{-(i\omega_{12}+\frac{1}{2}\gamma_{12})t_0}\Phi\{(i\omega_{12} + \tfrac{1}{2}\gamma_{12})T_c\}\sigma_{12}^0 \tag{37b}$$

characterize the atomic beam at the point of entry into the field. Equations (36) and (37a, b) show that, if the process of creation leads to an incoherent atomic beam ($\sigma_{12}^0 = 0$), then the interaction with a single rf field cannot produce coherence either; that is, ρ is a diagonal matrix.

Figure 3.3 illustrates the effect of the rf field phase on occupation probability *before* the ensemble average (35) is taken for the $4^2S_{1/2}(10)$–$4_2P_{1/2}(00)$ transition. With both states initially populated, the lineshapes computed with different values of δ are asymmetrical and show a marked variation from the usually expected Lorentzian-like profile. The phase-averaged lineshape (ρ_{11} vs. $\omega/2\pi$), however, is symmetrical and in good

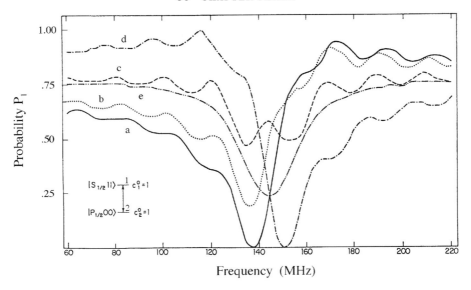

Figure 3.3 Effect of the rf phase δ on the two-state GRF probability $P_1 = |\langle 1 \mid \Psi \rangle|^2$ with initial amplitudes $c_1^0 = c_2^0 = 1$. (a) $\delta = 0$, (b) $\delta = \pi/4$, (c) $\delta = \pi/2$, (d) $\delta = \pi$, (e) phase-averaged spectrum. Parameters: $E_0 = 1$ V/cm, $t = 50$ ns, $\omega_{12}/2\pi = 144$ MHz, $\gamma_1 = 4.35 \times 10^6$ s^{-1}, $\gamma_2 = 80.6 \times 10^6$ s^{-1}.

agreement with experimentally observed lineshapes for this type of transition.

It is worth noting now—although the matter will be discussed more fully in chapter 7—that the phase of the rf field is not always a distributed quantity, but can be well defined. Such a situation could arise, for example, if the beam were pulsed so that atoms always arrived at a definite point in the oscillation cycle. Also, in the case of multiple rf fields, the relative phase of the various oscillators can be held to a fixed value; this is the basis for the Ramsey method of separated oscillating fields[7] to narrow resonance lineshapes.

3.5 THREE-STATE TRANSITIONS

An oscillating field interacting with hydrogen atoms in the absence of static fields can couple the state $|LJFF\rangle$ of one fine structure level to the nondegenerate states $|L+1\ J+1\ F\ F\rangle$ and $|L+1\ J+1\ F+1\ F\rangle$ of the adjacent higher-lying fine structure level to produce an irreducible three-level system. In this case, by setting $\lambda_1 = \Omega_{13}$, $\lambda_2 = \Omega_{23}$, and $\lambda_3 = 0$, one can reduce the Π matrix to time-independent form

$$
\Pi^{(3)} = \begin{pmatrix} -\left(\frac{1}{2}\gamma_1 - i\Omega_{13}\right) & 0 & -iV_{13}e^{-i\delta} \\ 0 & -\left(\frac{1}{2}\gamma_2 - i\Omega_{23}\right) & -iV_{23}e^{-i\delta} \\ -iV_{13}e^{i\delta} & -iV_{23}e^{i\delta} & -\frac{1}{2}\gamma_3 \end{pmatrix}. \quad (38)
$$

Determination of the eigenvalues and eigenvectors of $\Pi^{(3)}$ from which the matrices ε and C are constructed requires solution of a cubic algebraic equation, a straightforward but tedious calculation left to the appendix. The components of the state vector (30) then take the explicit form

$$
c_1(t) = e^{-i(\omega+\omega_3)(t-t_0)}
$$
$$
\times \left[M_{11}c_1(t_0) + M_{12}c_2(t_0) + e^{-i\omega t_0 - i\delta}M_{13}c_3(t_0) \right], \quad (39a)
$$

$$
c_2(t) = e^{-i(\omega+\omega_3)(t-t_0)}
$$
$$
\times \left[M_{21}c_1(t_0) + M_{22}c_2(t_0) + e^{-i\omega t_0 - i\delta}M_{23}c_3(t_0) \right], \quad (39b)
$$

$$
c_3(t) = e^{-i\omega_3(t-t_0)+i\omega t_0 + i\delta}
$$
$$
\times \left[M_{31}c_1(t_0) + M_{32}c_2(t_0) + e^{-i\omega t_0 - i\delta}M_{33}c_3(t_0) \right], \quad (39c)
$$

in which the elements of the matrix M are defined by

$$
M_{\mu\nu} = \sum_{\kappa=1}^{3} e^{\varepsilon_\kappa(t-t_0)}[C_{\mu\kappa}C_{\kappa\nu}^{-1}]_{\delta=0}. \quad (40)
$$

In figure 3.4 the GRF occupation probabilities $|c_\mu|^2$ from Eqs. (39a–c) are compared with results of exact numerical integration of the Schrödinger equation for transitions between $4^2P_{3/2}(22)$, $4^2D_{5/2}(32)$, and $4^2D_{5/2}(22)$ states. As initial conditions, the two D states are taken to be equally populated ($c_1^0 = c_2^0 = 1$) and the shorter-lived P state to be unpopulated ($c_3^0 = 0$)—a choice that results in occupation probabilities independent of the phase of the rf field. The agreement between the two calculational methods is quite satisfactory even for relatively

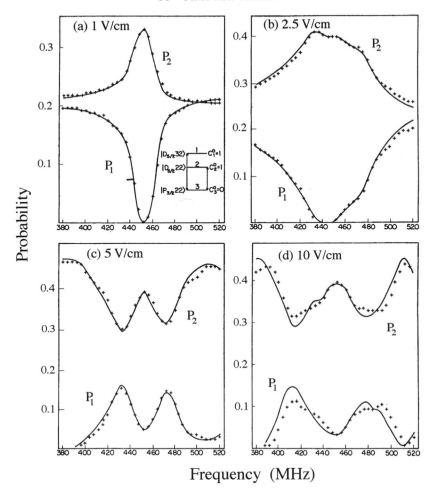

Figure 3.4 Comparison of probabilities $P_1 = |\langle 1 \mid \Psi \rangle|^2$ and $P_2 = |\langle 2 \mid \Psi \rangle|^2$ determined by three-state GRF theory (crosses) and exact numerical integration (solid lines) with initial amplitudes $c_1^0 = c_2^0 = 1$, $c_3^0 = 0$. Parameters: $t = 50$ ns, $\omega_{13}/2\pi = 454$ MHz, $\omega_{23}/2\pi = 453$ MH, $\gamma_1 = \gamma_2 = 27.4 \times 10^6$ s^{-1}, $\gamma_3 = 80.6 \times 10^6$ s^{-1}.

strong electric fields; a not atypical field strength employed in hydrogen electric resonance experiments is approximately 2 V/cm peak to peak.

Unlike the case of two coupled states (in the rotating-wave approximation), three-level couplings lead to asymmetric GRF lineshapes that can peak at different frequencies. In addition, the occupation or transition probabilities in a three-state system depend on the sign, as well as magnitude, of the coupling elements $V_{\mu\nu}$, whereas only $|V_{12}|^2$ enters the theory of a two-state system. Thus, as a consequence of the negative sign of $V_{23} \propto \langle 4^2 D_{5/2} 22 | x | 4^2 P_{3/2} 22 \rangle$, the population of the state $D_{5/2}(22)$ is increased at the expense of the state $D_{5/2}(32)$ although there is no direct coupling between them. Atoms are pumped from the latter to the former via $P_{3/2}(22)$ as an intermediary, a process that obviously cannot occur when only two coupled levels are involved. It is thus to be expected that the $(1 \leftrightarrow 3)$ and $(2 \leftrightarrow 3)$ transition probabilities for the two pairs of separately coupled levels can differ markedly from those predicted by three-level theory—and figure 3.5 indeed shows this to be the case.

The effect of interaction time on lineshape is illustrated in figure 3.6 for each of the three states. As in the case of two coupled states, the three-state resonance profiles narrow with increasing time spent in the rf field, and approach a limit set by the spontaneous decay rate. For times much longer than the state lifetime $\tau_\mu = \gamma_\mu^{-1}$ the distortion and asymmetry increase and the apparent resonance maximum may also shift.

Averaging bilinear products of the amplitudes given by Eqs. (39) and (40) over the field phase yields the density matrix of the beam. The diagonal elements, with which we will be primarily concerned, take the form

$$\rho_{\mu\mu}(t) = \sum_{\nu=1}^{3} |M_{\mu\nu}|^2 \sigma_{\nu\nu}(t_0) + 2 \operatorname{Re}\{M_{\mu 1} M_{\mu 2}^* \sigma_{12}(t_0)\}, \qquad (41)$$

where $\operatorname{Re}\{z\}$ signifies the real part of z. Because states $|1\rangle$ and $|2\rangle$ are different hyperfine states of the same fine structure level, they can be populated coherently (as will be demonstrated later) even if atoms in different orbital states were produced incoherently. Thus, interference terms enter all the diagonal elements, a feature not found in the two-level theory. The effect of the interference term is illustrated in figure 3.7 for the state $4^2 D_{3/2}(22)$ coupled to $4^2 F_{5/2}(32)$ and $4^2 F_{5/2}(22)$. In general, the lineshape (diagonal elements of ρ as a function of frequency) of the fully coherent three-level system is oscillatory with rapid variation in the vicinity of resonance. The profile smoothes out with diminishing coherence σ_{12}^0.

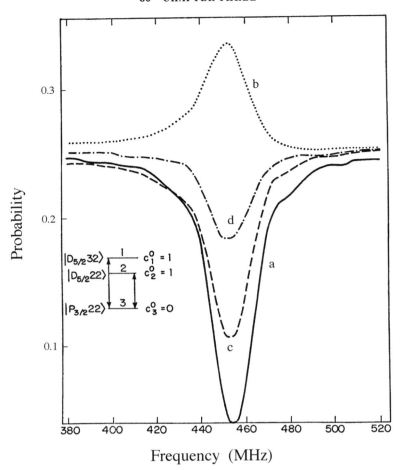

Figure 3.5 Three-state GRF probabilities (a) $P_1 = |\langle 1 \mid \Psi \rangle|^2$, (b) $P_2 = |\langle 2 \mid \Psi \rangle|^2$ and two-state GRF probabilities (c) P_1, (d) P_2. $E_0 = 1$ V/cm, and other parameters are the same as for figure 3.4. The difference between lineshapes (b) and (d) is due principally to the dependence of three-state amplitudes on the sign of the dipole matrix elements.

3.6 FOUR-STATE TRANSITIONS

Proceeding in a manner similar to that of the previous section, we obtain the time-independent interaction Hamiltonian

$$\Pi^{(4)} = \begin{pmatrix} -(\frac{1}{2}\gamma_1 + i\omega_{12}) & 0 & -iV_{13}e^{-i\delta} & 0 \\ 0 & -\frac{1}{2}\gamma_2 & -iV_{23}e^{-i\delta} & -iV_{24}e^{-i\delta} \\ -iV_{13}e^{i\delta} & -iV_{23}e^{i\delta} & -(\frac{1}{2}\gamma_3 + i\Omega_{13}) & 0 \\ 0 & -iV_{24}e^{i\delta} & 0 & -(\frac{1}{2}\gamma_4 + i\Omega_{24}) \end{pmatrix} \quad (42)$$

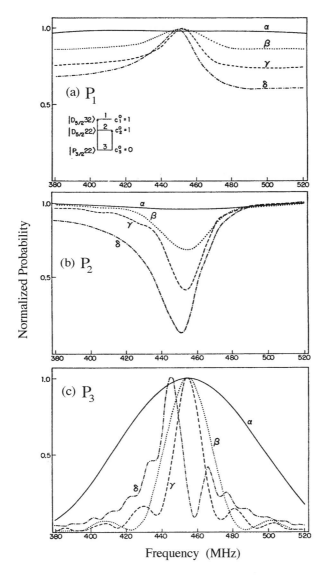

Figure 3.6 Variation of GRF three-state probabilities $P_i = |\langle i \mid \Psi \rangle|^2$ with interaction time (in ns): (α) 10, (β) 30, (γ) 50, (δ) 100. $E_0 = 1$ V/cm; other parameters are the same as for figure 3.4.

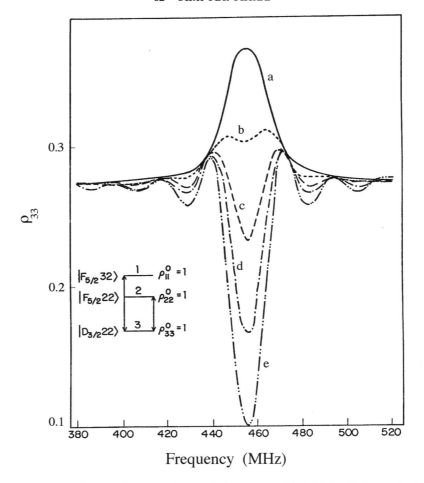

Figure 3.7 Variation of occupation probability ρ_{33} with initial off-diagonal element $\sigma_{12}^0 =$ (a) 0, (b) 0.25, (c) 0.5, (d) 0.75, (e) 1.00. Interaction time $t = 50$ ns, $\omega_{13}/2\pi = 456$ MHz, $\omega_{23}/2\pi = 455$ MHz, $\gamma_1 = \gamma_2 = 13.7 \times 10^6$ s^{-1}, $\gamma_3 = 27.4 \times 10^6$ s^{-1}.

by setting $\lambda_1 = -\omega_{12}$, $\lambda_2 = 0$, $\lambda_3 = -\Omega_{23}$, and $\lambda_4 = -\Omega_{24}$. Determination of the eigenvalues and eigenvectors of this four-dimensional system requires solution of a quartic algebraic equation, the details of which are again left to the appendix. Once accomplished, the wavefunction (30) yields state amplitudes of the form

$$c_1(t) = e^{-i\omega_2(t-t_0)}\big[M_{11}c_1(t_0) + M_{12}c_2(t_0)$$
$$+ e^{-i\omega t_0 - i\delta}\{M_{13}c_3(t_0) + M_{14}c_4(t_0)\}\big], \qquad (43a)$$

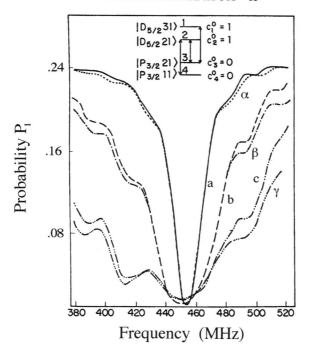

Figure 3.8 Comparison of occupation probability P_1 computed by four-state GRF theory (α, β, γ) and exact numerical integration (a, b, c). E_0 (in V/cm) = (a, α) 1, (b, β) 2.5, (c, γ) 5. Parameters $t = 50$ ns, $\omega_{13}/2\pi = 454$, $\omega_{23}/2\pi = 453$, and $\omega_{24}/2\pi = 456$ MHz, $\gamma_1 = \gamma_2 = 27.4 \times 10^6$ s^{-1}, $\gamma_3 = \gamma_4 = 80.6 \times 10^6$ s^{-1}.

$$c_2(t) = e^{-i\omega_2(t-t_0)}\big[M_{21}c_1(t_0) + M_{22}c_2(t_0)$$
$$+ e^{-i\omega t_0 - i\delta}\{M_{23}c_3(t_0) + M_{24}c_4(t_0)\}\big], \qquad (43b)$$

$$c_3(t) = e^{-i\omega_2(t-t_0)+i\omega t+i\delta}\big[M_{31}c_1(t_0) + M_{32}c_2(t_0)$$
$$+ e^{-i\omega t_0 - i\delta}\{M_{33}c_3(t_0) + M_{34}c_4(t_0)\}\big], \quad (43c)$$

$$c_4(t) = e^{-i\omega_2(t-t_0)+i\omega t+i\delta}\big[M_{41}c_1(t_0) + M_{42}c_2(t_0)$$
$$+ e^{-i\omega t_0 - i\delta}\{M_{43}c_3(t_0) + M_{44}c_4(t_0)\}\big], \quad (43d)$$

in which the elements $M_{\mu\nu}$ are given by the summation (40) with upper limit extended to 4.

A comparison of resonance lineshapes computed by GRF theory and by numerical integration is shown in figure 3.8 for the coupled states

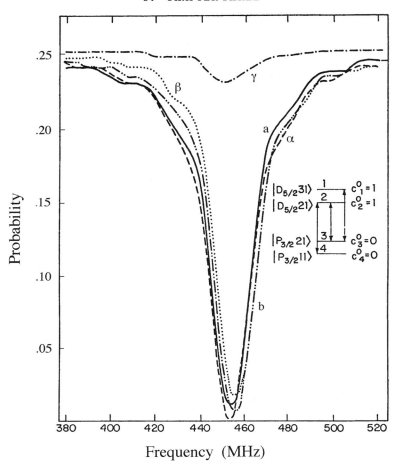

Figure 3.9 Four-state GRF calculation of (a) $P_1 = |\langle 1 | \Psi \rangle|^2$ and (b) $P_2 = |\langle 2 | \Psi \rangle|^2$; two-state GRF calculation of (α) P_1 (transition $1 \leftrightarrow 3$), (β) P_2 (transition $2 \leftrightarrow 4$), (γ) P_2 (transition $2 \leftrightarrow 3$). $E_0 = 1$ V/cm; other parameters the same as for figure 3.8.

$4^2D_{5/2}(31)$, $4^2D_{5/2}(21)$, $4^2P_{3/2}(21)$, $4^2P_{3/2}(11)$. Agreement is reasonably good although deviation in the wings occurs for field strengths large enough to broaden and distort the profiles severely. Nevertheless, the four-level (and three-level) theory reproduces the major lineshape asymmetries predicted by the exact numerical solution.

An interesting point is that for all sets of four coupled levels in hydrogen, the interaction matrix element $|V_{23}|$ is considerably smaller than

$|V_{13}|$ or $|V_{24}|$; for example, in the $n = 4$ manifold, V_{23}^2 is at least a factor of 10 smaller than V_{13}^2 or V_{24}^2. In the limit that $V_{23} \to 0$, the four coupled equations (28) reduce to two pairs of two coupled equations. One would therefore expect the four-level transition probabilities in hydrogen to resemble somewhat closely those obtained by treating each transition by the two-level theory, a feature not expected in the three-level case. In figure 3.9 this behavior is confirmed for the four-level $D_{5/2}$–$P_{3/2}$ system.

The diagonal density matrix elements that follow from the set of amplitudes (43a–d) take the form

$$\rho_{\mu\mu}(t) = \sum_{\nu=1}^{4} |M_{\mu\nu}|^2 \sigma_{\nu\nu}(t_0) + 2\,\mathrm{Re}\{M_{\mu 1} M_{\mu 2}^* \sigma_{12}(t_0)\}$$
$$+ 2\,\mathrm{Re}\{M_{\mu 3} M_{\mu 4}^* \sigma_{34}(t_0)\}, \tag{44}$$

which now exhibits interference terms from both the upper and lower pair of states. The effects of different values of $\sigma_{12}(t_0)$ and $\sigma_{34}(t_0)$ on the occupation probability of the state $4F_{5/2}(31)$ coupled directly to $4D_{3/2}(21)$ and indirectly to $4F_{5/2}(21)$ and $4D_{3/2}(11)$ are illustrated in figure 3.10. All states are taken to be equally populated at the outset. The occupation probability of the fully coherent system ($\sigma_{12}^0 = \sigma_{34}^0 = 1$) is highly oscillatory with large variation between 0 and ~ 0.8 in the vicinity of resonance. Phase averaging progressively removes the major contribution to the oscillatory interference terms leaving nonundulatory lineshapes only weakly influenced by the off-diagonal density matrix elements σ_{12}^0 and σ_{34}^0. This feature is exhibited as well by the occupation probabilities $\rho_{\mu\mu}$ of the other three levels $\mu = 2, 3, 4$.

3.7 NUMERICAL SOLUTION OF THE N-STATE SYSTEM

It is often essential to be able to solve the coupled N-state Schrödinger equation exactly by numerical means if only as a check on analytical solutions obtained by various approximations. The term "exact" recognizes, of course, that any numerical procedure returns a result of finite precision generally limited by computation time and round-off error.

The numerical calculations of the two-, three-, and four-level GRF solutions in this book were programmed in FORTRAN on a large main-

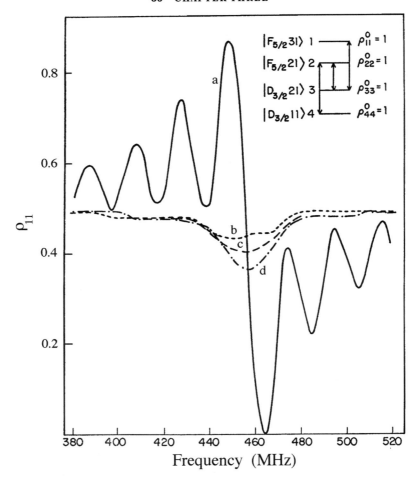

Figure 3.10 Variation of occupation probability ρ_{11} with initial off-diagonal elements σ_{12}^0, σ_{34}^0. (a) Complete coherence (no phase average); $\sigma_{12}^0 = \sigma_{34}^0 = $ (b) 1.0, (c) 0.5, (d) 0. Parameters are $t = 50$ ns, $\omega_{13}/2\pi = 456$, $\omega_{23}/2\pi = 455$, and $\omega_{24}/2\pi = 457$ MHz, $\gamma_1 = \gamma_2 = 13.7 \times 10^6$ s^{-1}, $\gamma_3 = \gamma_4 = 27.4 \times 10^6$ s^{-1}.

frame computer at a time when desk-top computers and powerful mathematical applications like MAPLE, Mathematica, and others were not available. Nevertheless, the numerical procedure, a generalization of Euler's method[8] to a set of simultaneous equations, is still useful. It has the advantage of being simple, easy to program, and especially suitable for first-order linear differential equations such as result from the time-dependent Schrödinger equation. In the case of a multilevel atom interacting with a single oscillating field, we start with the equation of motion (20), here rewritten as

$$\frac{da_\mu(t)}{dt} = -\tfrac{1}{2}\gamma_\mu a_\mu(t) - i\sum_{\nu=1}^{N} V_{\mu\nu} e_{\mu\nu}(t) a_\nu(t) \qquad (45)$$

with

$$e_{\mu\nu}(t) = e^{i(\Omega'_{\mu\nu}t+\delta)} + e^{-i(\Omega_{\mu\nu}t+\delta)}. \qquad (46)$$

Although the amplitudes on the right side of Eq. (45) are unknown, the equation can be integrated recursively by the algorithm

$$a_\mu(t_k) = a_\mu(t_{k-1}) - \sum_{\nu=1}^{N}\{\tfrac{1}{2}\gamma_\nu\delta_{\mu\nu} + iV_{\mu\nu}e_{\mu\nu}(t_{k-1})\}a_\nu(t_{k-1})\,\delta t$$
$$(k = 1, 2, \ldots, n) \quad (47)$$

starting from a set of known initial conditions $\{a_\mu(t_0) \equiv a_\mu^0\}$. The step size δt or number of intervals n, such that $n\delta t = t$, is chosen to obtain satisfactory accuracy within the computation time available (which on a personal computer is no longer a costly limitation).

Euler's method is said to be of first order because the truncation error varies linearly with step size δt. For physical processes in which only a few levels are closely coupled, as, for example, the rf coupling of hydrogen fine structure levels, this technique is quite satisfactory. For larger-scale problems, however, a more efficient algorithm would be required, such as Heun's method which is second order in δt, or the widely used classical Runge-Kutta method which is fourth order in δt.

The numerical procedure outlined above worked well for computing GRF state amplitudes and occupation probabilities, but was not employed for constructing density matrix elements because of the impracticality in implementing the requisite phase averages numerically.

3.8 COUPLING ELEMENTS $V_{\mu\nu}$

For hydrogen states with the axis of quantization defined by the rf electric field (here taken to be the x-axis), the electric dipole matrix elements $V_{\mu\nu}$ of Eq. (18) reduce to

$$V_{\mu\nu} = -eE_0 x_{\mu\nu}. \qquad (48)$$

E_0 is the magnitude of the electric field, and the coordinate matrix element $x_{\mu\nu}$, expressed in terms of Wigner 3-J and 6-J symbols[9] for ease of computation, takes the somewhat formidable form

$$\langle nL_1J_1F_1M_1|x|nL_2J_2F_2M_2\rangle$$
$$= (-)^{J_1+J_2-M_1} \times \tfrac{3}{2}n\{(2F_1+1)(2F_2+1)(2J_1+1)(2J_2+1)$$
$$\times (2L_1+1)(2L_2+1)(n^2-L_>^2)\}^{1/2}$$
$$\times \begin{Bmatrix} J_1 & J_2 & 1 \\ F_2 & F_1 & \tfrac{1}{2} \end{Bmatrix} \begin{Bmatrix} L_1 & L_2 & 1 \\ J_2 & J_1 & \tfrac{1}{2} \end{Bmatrix}$$
$$\times \begin{pmatrix} F_2 & 1 & F_1 \\ M_2 & 0 & -M_1 \end{pmatrix} \begin{pmatrix} L_1 & 1 & L_2 \\ 0 & 0 & 0 \end{pmatrix}. \tag{49}$$

The symbol $L_>$ in Eq. (49) stands for the greater of L_1 or L_2.

Although to delve into the intricacies of angular momentum theory is beyond the scope of this book, a brief account of the purpose and properties of the Wigner symbols may be helpful in elucidating the structure of the above equation. It is often the case in quantum physics that one must calculate an expectation value or transition matrix element of an operator that acts on only one part of a composite system. Such is the case here; the electric dipole operator $\mu_E = -e\mathbf{r}$ acts on the orbital states $|LM_L\rangle$ that are coupled to electron and nuclear spin states to create the coupled angular momentum (c.a.m.) basis of field-free eigenstates of the (nonrelativistic) hydrogen Hamiltonian. Evaluation of $x_{\mu\nu}$ therefore requires a decomposition of the c.a.m. states $\langle\mu|$ and $|\nu\rangle$ into direct products of the original uncoupled bases so that components of the coordinate operator $\mathbf{r}$ can be "sandwiched" between orbital (LM_L) states only.

The coupling of two sets of angular momentum states $|a\alpha\rangle$ and $|b\beta\rangle$ to form a composite system $|c\gamma\rangle$ is effected by the vector coupling or Clebsch-Gordan (CG) coefficients $\langle ab\alpha\beta|c -\gamma\rangle$ (or $\langle c -\gamma|ab\alpha\beta\rangle$ for combining the adjoint vectors) according to the relation

$$|c\gamma\rangle = \sum_{\alpha,\beta}\langle ab\alpha\beta|c -\gamma\rangle|a\alpha\rangle|b\beta\rangle, \tag{50}$$

introduced in most quantum mechanics texts. For given angular momentum quantum numbers a and b, the allowed values of c satisfy the "triangular condition"

$$a + b \geq c \geq |a - b|, \qquad (51)$$

whereby c, which ranges from the lower to upper limit in integer steps, can form the third leg of a triangle whose first two legs are of lengths a and b. Correspondingly, the conservation of angular momentum requires that the three magnetic quantum numbers sum to zero,

$$\alpha + \beta + \gamma = 0, \qquad (52)$$

otherwise the CG coefficient vanishes identically.

The Wigner 3-*J* symbol $\begin{pmatrix} a & b & c \\ \alpha & \beta & \gamma \end{pmatrix}$, defined by

$$\begin{pmatrix} a & b & c \\ \alpha & \beta & \gamma \end{pmatrix} \sqrt{2c+1} = (-1)^{a-b-\gamma} \langle ab\alpha\beta \mid c -\gamma \rangle, \qquad (53)$$

is often more useful than the CG coefficient when substantial algebraic manipulation is called for because of its high degree of symmetry. For example, it can be reexpressed in the form of a 3×3 matrix whose symmetry properties yield 72 equivalent symbols. Among the most useful properties of the 3-*J* symbol are that (1) it is invariant under an even (cyclic) permutation of its columns, (2) it incurs a phase factor $(-1)^{a+b+c}$ for an odd permutation of its columns, and (3) it incurs the preceding phase factor under a sign change of all magnetic quantum numbers, $(\alpha, \beta, \gamma) \rightarrow (-\alpha, -\beta, -\gamma)$. In this way the use of 3-*J* symbols avoids the unsymmetrical surds and phases that appear in the corresponding symmetry relations for the CG coefficients. All three angular momenta (a, b, c) are displayed and manipulated on equal footing, a feature ideally suited for coupling and decoupling basis states.

Although it will not be demonstrated here, 3-*J* symbols satisfy two orthogonality relations highly useful in the reduction of complex expressions of matrix elements with multiple sums over quantum numbers,

$$\sum_{\alpha,\beta}(2c+1)\begin{pmatrix} a & b & c \\ \alpha & \beta & \gamma \end{pmatrix}\begin{pmatrix} a & b & c' \\ \alpha & \beta & \gamma' \end{pmatrix} = \delta_{cc'}\delta_{\gamma\gamma'}, \quad (54a)$$

$$\sum_{c}(2c+1)\begin{pmatrix} a & b & c \\ \alpha & \beta & \gamma \end{pmatrix}\begin{pmatrix} a & b & c \\ \alpha' & \beta' & \gamma \end{pmatrix} = \delta_{\alpha\alpha'}\delta_{\beta\beta'}. \quad (54b)$$

From the defining relation (53) and a general expression for CG coefficients first derived by Racah, there follows the formula

$$\begin{pmatrix} a & b & c \\ \alpha & \beta & \gamma \end{pmatrix} = (-1)^{a-b-\gamma}\delta_{\alpha+\beta,\gamma}\Delta(abc)$$

$$\times \left[(a+\alpha)!(a-\alpha)!(b+\beta)!(b-\beta)!(c+\gamma)!(c-\gamma)!\right]^{1/2}$$

$$\times \sum_{\kappa}(-1)^{\kappa}\left[(a-\alpha-\kappa)!(c-b+\alpha+\kappa)!\right.$$

$$\times (b+\beta-\kappa)!(c-a-\beta+\kappa)!\kappa!$$

$$\times (a+b-c-\kappa)!\Big]^{-1},$$

$$(55a)$$

which can be readily programmed for electronic computation. The triangle function $\Delta(abc)$ is defined by

$$\Delta(abc) = \left[\frac{(a+b-c)!(a+c-b)!(b+c-a)!}{(a+b+c+1)!}\right]^{1/2}, \quad (55b)$$

and the index κ of the summation in (55a) runs over all integer values that do not lead to a negative factorial.

Although the coupling of two angular momenta leads in effect to a unique set of c.a.m. basis states, the coupling of three angular momenta can be performed in three different ways leading to different bases. For example, when coupling $|a\alpha\rangle$, $|b\beta\rangle$, $|c\gamma\rangle$ one can first combine a and b to produce J_{ab} to which c is then added vectorially. Alternatively, one can combine first a and c to produce J_{ac}, or b and c to produce J_{bc}. The three sets of intermediate basis states, which are eigenstates of different sets of operators, are not independent but are connected by a linear transformation like the following:

$$|(ab)J_{ab}, c; JM\rangle = \sum_{J_{bc}} \sqrt{(2J_{ab} + 1)(2J_{bc} + 1)}$$

$$\times W(abJc; J_{ab}J_{bc})|a, (b, c)J_{bc}; JM\rangle, \quad (56)$$

in which the transformation coefficients, known as the Racah W-functions, play a role analogous to that of the CG coefficients.

Parallel to his treatment of the vector-addition coefficients expressed as a 2×3 array, Wigner defined a 6-J symbol differing from the Racah function only by a phase,

$$\begin{Bmatrix} a & b & e \\ d & c & f \end{Bmatrix} = (-1)^{a+b+c+d} W(abcd; ef). \quad (57)$$

Like the 3-J symbol, the 6-J symbol satisfies a number of symmetry relations, giving rise to 144 equivalent coefficients. For example, it is invariant under the interchange of any two columns as well as the exchange of upper and lower elements in each of any two columns. From a general formula for W first derived by Racah as a scalar contraction of the product of four CG coefficients, there follows the explicit expression

$$\begin{Bmatrix} a & b & e \\ d & c & f \end{Bmatrix}$$

$$= (-1)^{a+b+c+d} \Delta(abe)\Delta(acf)\Delta(bdf)\Delta(cde) \times \sum_\kappa (-1)^\kappa$$

$$\times \frac{(a+b+c+d+1-\kappa)!}{[\kappa!(e+f-a-d+\kappa)!(e+f-b-c+\kappa)!(a+b-e-\kappa)!}{\times(c+d-e-\kappa)!(a+c-f-\kappa)!(b+d-f-\kappa)!]}.$$

$$(58)$$

Often it is useful simply to know whether a dipole matrix element coupling two states vanishes—and this can be ascertained at a glance from the fact that the four sets of angular momenta (abe), (acf), (dbf), (dce) must satisfy the triangular condition. It therefore follows from the properties of the 3-J and 6-J symbols in Eq. (49) that only those electric dipole transitions occur that satisfy the selection rules

$$|\Delta L| = 1, \qquad |\Delta J| = 0, 1, \quad \text{and } |\Delta F| = 0, 1 \qquad (59)$$

(with no transitions between $F_1 = 0$ and $F_2 = 0$). Moreover, one can also show that

$$\langle nL_1J_1F_1M_1|x|nL_2J_2F_2M_2\rangle = (-)^{F_1+F_2+1}\langle nL_1J_1F_1\bar{M}_1|x|nL_2J_2F_2\bar{M}_2\rangle \qquad (60)$$

for transitions between states of negative magnetic quantum number (designated by an overbar).

APPENDIX: EIGENVALUES AND EIGENVECTORS OF THREE- AND FOUR-STATE SYSTEMS

The eigenvalue problems presented by matrices $\Pi^{(3)}$ and $\Pi^{(4)}$ require the solution of a cubic and quartic equation, respectively, for which suitable methods can be found in standard works on the theory of equations.[10] The derivation of these solutions will be left to the reference, but the resulting analytical expressions for the eigenvalues and eigenvectors will be summarized below.

3.A.1 Three-State System

The cubic secular equation of the matrix

$$\Pi^{(3)} = \begin{pmatrix} \Gamma_1 & 0 & -iV_{13}e^{-i\delta} \\ 0 & \Gamma_2 & -iV_{23}e^{-i\delta} \\ -iV_{13}e^{i\delta} & -iV_{23}e^{i\delta} & \Gamma_3 \end{pmatrix} \qquad (A1)$$

corresponding to Eq. (38) can be solved by a procedure attributed to Cardan ($\sim$ 1545). Expressed trigonometrically, this leads to the eigenvalues

$$\varepsilon_1 = -\tfrac{1}{3}b + n\cos\left(\theta + \frac{4\pi}{3}\right), \qquad (A2a)$$

$$\varepsilon_2 = -\tfrac{1}{3}b + n\cos\left(\theta + \frac{2\pi}{3}\right), \tag{A2b}$$

$$\varepsilon_3 = -\tfrac{1}{3}b + n\cos\theta, \tag{A2c}$$

where

$$\theta = \cos^{-1}(-4q/n^3), \tag{A3}$$

$$n = \sqrt{-\tfrac{4}{3}(c - \tfrac{1}{3}b^2)}, \tag{A4}$$

$$q = d - \tfrac{1}{3}bc + \tfrac{2}{27}b^3, \tag{A5}$$

and

$$b = \Gamma_1 + \Gamma_2 + \Gamma_3, \tag{A6}$$

$$c = \Gamma_1\Gamma_2 + \Gamma_1\Gamma_3 + \Gamma_2\Gamma_3 + V_{13}^2 + V_{23}^2, \tag{A7}$$

$$d = \Gamma_1\Gamma_2\Gamma_3 + \Gamma_1 V_{23}^2 + \Gamma_2 V_{13}^2. \tag{A8}$$

The matrix of eigenvectors C can be written in the form

$$C(\delta) = \begin{pmatrix} 1 & k_{12} & k_{13}e^{-i\delta} \\ k_{21} & 1 & k_{23}e^{-i\delta} \\ k_{31}e^{i\delta} & k_{32}e^{i\delta} & 1 \end{pmatrix} \tag{A9}$$

with elements

$$k_{12} = \frac{-V_{13}V_{23}}{(\varepsilon_2 + \Gamma_1)(\varepsilon_2 + \Gamma_3) + V_{13}^2}, \tag{A10a}$$

$$k_{13} = \frac{-iV_{13}}{(\varepsilon_3 + \Gamma_1)}, \tag{A10b}$$

$$k_{21} = \frac{-V_{13}V_{23}}{(\varepsilon_1 + \Gamma_2)(\varepsilon_1 + \Gamma_3) + V_{23}^2}, \tag{A10c}$$

$$k_{23} = \frac{-iV_{23}}{(\varepsilon_3 + \Gamma_2)}, \tag{A10d}$$

$$k_{31} = \frac{-iV_{13}(\varepsilon_1 + \Gamma_2)}{(\varepsilon_1 + \Gamma_2)(\varepsilon_1 + \Gamma_3) + V_{23}^2}, \tag{A10e}$$

$$k_{32} = \frac{-iV_{23}(\varepsilon_2 + \Gamma_1)}{(\varepsilon_2 + \Gamma_1)(\varepsilon_2 + \Gamma_3) + V_{13}^2}, \tag{A10f}$$

3.A.2 *Four-State System*

By a method of solution attributable to Ferrari (1522–1565), the general algebraic equation of degree four can be factored into the product of two quadratic equations provided that a certain expression of degree three (the resolvent cubic) vanishes. In this way the quartic secular equation of the matrix

$$\Pi^{(4)} = \begin{pmatrix} \Gamma_1 & 0 & -iV_{13}e^{-i\delta} & 0 \\ 0 & \Gamma_2 & -iV_{23}e^{-i\delta} & -iV_{24}e^{-i\delta} \\ -iV_{13}e^{i\delta} & -iV_{23}e^{i\delta} & \Gamma_3 & 0 \\ 0 & -iV_{24}e^{i\delta} & 0 & \Gamma_4 \end{pmatrix} \quad \text{(A11)}$$

corresponding to Eq. (42) can be shown to lead to the eigenvalues

$$\varepsilon_1 = -\tfrac{1}{2}(\tfrac{1}{2}b - m) + \tfrac{1}{2}\sqrt{(\tfrac{1}{2}b - m)^2 - 4(\tfrac{1}{2}y_j - n)}, \quad \text{(A12a)}$$

$$\varepsilon_2 = -\tfrac{1}{2}(\tfrac{1}{2}b - m) - \tfrac{1}{2}\sqrt{(\tfrac{1}{2}b - m)^2 - 4(\tfrac{1}{2}y_j - n)}, \quad \text{(A12b)}$$

$$\varepsilon_3 = -\tfrac{1}{2}(\tfrac{1}{2}b + m) + \tfrac{1}{2}\sqrt{(\tfrac{1}{2}b + m)^2 - 4(\tfrac{1}{2}y_j + n)}, \quad \text{(A12c)}$$

$$\varepsilon_4 = -\tfrac{1}{2}(\tfrac{1}{2}b + m) - \tfrac{1}{2}\sqrt{(\tfrac{1}{2}b + m)^2 - 4(\tfrac{1}{2}y_j + n)}, \quad \text{(A12d)}$$

in which

$$m = \sqrt{\tfrac{1}{4}b^2 - c + y_j}, \quad \text{(A13)}$$

$$n = \frac{\tfrac{1}{2}by_j - d}{2m}, \quad \text{(A14)}$$

and y_j ($j = 1, 2, 3$) is a root of the cubic equation

$$y^3 - cy^2 + (bd - 4c)y - (b^2e - 4ce + d^2) = 0. \quad \text{(A15)}$$

The lettered coefficients are

$$b = \Gamma_1 + \Gamma_2 + \Gamma_3 + \Gamma_4, \quad \text{(A16a)}$$

$$c = \Gamma_1\Gamma_2 + \Gamma_1\Gamma_3 + \Gamma_1\Gamma_4 + \Gamma_2\Gamma_3 + \Gamma_2\Gamma_4 + \Gamma_3\Gamma_4 \\ + V_{13}^2 + V_{23}^2 + V_{24}^2, \quad \text{(A16b)}$$

$$d = \Gamma_1\Gamma_2\Gamma_3 + \Gamma_1\Gamma_2\Gamma_4 + \Gamma_1\Gamma_3\Gamma_4 + \Gamma_2\Gamma_3\Gamma_4$$
$$+ (\Gamma_1 + \Gamma_4)V_{23}^2 + (\Gamma_1 + \Gamma_3)V_{24}^2 + (\Gamma_2 + \Gamma_4)V_{13}^2, \qquad \text{(A16c)}$$

$$e = \Gamma_1\Gamma_2\Gamma_3\Gamma_4 + \Gamma_1\Gamma_4 V_{23}^2 + \Gamma_1\Gamma_3 V_{24}^2 + \Gamma_2\Gamma_4 V_{13}^2 + V_{13}^2 V_{24}^2. \quad \text{(A16d)}$$

The corresponding matrix of eigenvectors is

$$C(\delta) = \begin{pmatrix} 1 & k_{12} & k_{13}e^{-i\delta} & k_{14}e^{-i\delta} \\ k_{21} & 1 & k_{23}e^{-i\delta} & k_{24}e^{-i\delta} \\ k_{31}e^{i\delta} & k_{32}e^{i\delta} & 1 & k_{34} \\ k_{41}e^{i\delta} & k_{42}e^{i\delta} & k_{43} & 1 \end{pmatrix} \qquad \text{(A17)}$$

with components

$$k_{12} = \frac{-V_{13}V_{23}}{V_{13}^2 + (\varepsilon_2 + \Gamma_1)(\varepsilon_2 + \Gamma_3)}, \qquad \text{(A18a)}$$

$$k_{13} = \frac{-iV_{13}}{\varepsilon_3 + \Gamma_1}, \qquad \text{(A18b)}$$

$$k_{14} = \frac{iV_{13}V_{23}V_{24}}{(\varepsilon_4 + \Gamma_1)V_{23}^2 + (\varepsilon_4 + \Gamma_2)V_{13}^2 + (\varepsilon_4 + \Gamma_1)(\varepsilon_4 + \Gamma_2)(\varepsilon_4 + \Gamma_3)}, \qquad \text{(A18c)}$$

$$k_{21} = \frac{-V_{13}V_{23}(\varepsilon_1 + \Gamma_4)}{(\varepsilon_1 + \Gamma_3)V_{24}^2 + (\varepsilon_1 + \Gamma_4)V_{23}^2 + (\varepsilon_1 + \Gamma_2)(\varepsilon_1 + \Gamma_3)(\varepsilon_1 + \Gamma_4)}, \qquad \text{(A18d)}$$

$$k_{23} = \frac{-iV_{23}(\varepsilon_3 + \Gamma_4)}{V_{24}^2 + (\varepsilon_3 + \Gamma_2)(\varepsilon_3 + \Gamma_4)}, \qquad \text{(A18e)}$$

$$k_{24} = \frac{-iV_{24}[(\varepsilon_4 + \Gamma_1)(\varepsilon_4 + \Gamma_3) + V_{13}^2]}{(\varepsilon_4 + \Gamma_1)V_{23}^2 + (\varepsilon_4 + \Gamma_2)V_{13}^2 + (\varepsilon_4 + \Gamma_1)(\varepsilon_4 + \Gamma_2)(\varepsilon_4 + \Gamma_3)}, \qquad \text{(A18f)}$$

$$k_{31} = \frac{-iV_{13}[(\varepsilon_1 + \Gamma_2)(\varepsilon_1 + \Gamma_4) + V_{24}^2]}{(\varepsilon_1 + \Gamma_3)V_{24}^2 + (\varepsilon_1 + \Gamma_4)V_{23}^2 + (\varepsilon_1 + \Gamma_2)(\varepsilon_1 + \Gamma_3)(\varepsilon_1 + \Gamma_4)}, \qquad \text{(A18g)}$$

$$k_{32} = \frac{-iV_{23}(\varepsilon_2 + \Gamma_1)}{V_{13}^2 + (\varepsilon_2 + \Gamma_1)(\varepsilon_2 + \Gamma_3)}, \tag{A18h}$$

$$k_{34} = \frac{-V_{23}V_{24}(\varepsilon_4 + \Gamma_1)}{(\varepsilon_4 + \Gamma_1)V_{23}^2 + (\varepsilon_4 + \Gamma_2)V_{13}^2 + (\varepsilon_4 + \Gamma_1)(\varepsilon_4 + \Gamma_2)(\varepsilon_4 + \Gamma_3)}, \tag{A18i}$$

$$k_{41} = \frac{-iV_{13}V_{23}V_{24}}{(\varepsilon_1 + \Gamma_3)V_{24}^2 + (\varepsilon_1 + \Gamma_4)V_{23}^2 + (\varepsilon_1 + \Gamma_2)(\varepsilon_1 + \Gamma_3)(\varepsilon_1 + \Gamma_4)}, \tag{A18j}$$

$$k_{42} = \frac{-iV_{24}}{\varepsilon_2 + \Gamma_4}, \tag{A18k}$$

$$k_{43} = \frac{-V_{23}V_{24}}{V_{24}^2 + (\varepsilon_3 + \Gamma_2)(\varepsilon_3 + \Gamma_4)}. \tag{A18l}$$

NOTES

1. Two wavefunctions that differ only by a global phase factor lead to indistinguishable predictions of measurable quantities. If combined with a reference wave, the two wavefunctions may lead to different quantum interference effects, but in that case the phase is no longer global.

2. M. P. Silverman, "On Measurable Distinctions between Quantum Ensembles," in *New Techniques and Ideas in Quantum Measurement Theory*, ed. by D. M. Greenberger (New York Academy of Sciences, New York, 1986), pp. 292–303.

3. To simplify notation, quantum numbers that are the same for all states (such as $S = I = \frac{1}{2}$ for hydrogen) will be suppressed.

4. Equation (5) follows readily from the Schrödinger equation: $\mathcal{H}\Psi = i\partial\Psi/\partial t$. Multiply the latter on the right by $\Psi^\dagger$, multiply the Hermitian adjoint of the Schrödinger equation on the left by Ψ, subtract one from the other, and take an ensemble average.

5. M. P. Silverman, S. Haroche, and M. Gross, "General Theory of Laser-Induced Quantum Beats," Parts I and II, *Phys. Rev. A* **18** (1978):1507; 1517.

6. M. P. Silverman, *More than One Mystery: Explorations in Quantum Interference* (Springer, New York, 1995), chap. 5.

7. N. Ramsey, *Molecular Beams* (Oxford University Press, New York, 1963), pp. 124–134.

8. P. Henrici, *Elements of Numerical Analysis* (Wiley, New York, 1964), p. 267.

9. See D. M. Brink and G. R. Satchler, *Angular Momentum* (Oxford University Press, Oxford, 1968).

10. L. E. Dickson, *New First Course in the Theory of Equations* (Wiley, New York, 1939) pp. 42–55.

CHAPTER 4

Multiple-Quantum Transitions

4.1 THE QUANTIZED RADIOFREQUENCY FIELD

Until now we have regarded the rf field as a classical oscillating field external to a system of quantized atomic states. Such an approach is adequate for treatment of single-quantum transitions. There are situations, however, for which the description of the rf field and atom together as a single fully quantized system provides conceptual and practical advantages over the semiclassical formalism. One example is that of multiple-quantum (m.q.) transitions induced between two atomic states in which several photons together fulfill the requirements of energy and angular momentum conservation. Although semiclassical radiation theory is capable of treating such processes when the photon density is sufficiently large—and a derivation of the pertinent relations is given in appendix 4A—the quantized field formalism handles a broader range of problems and affords deeper physical insights that facilitate mathematical analysis.

It is useful to distinguish the m.q. electric-dipole processes of this chapter from those observed in magnetic-dipole transitions between states differing only in their magnetic quantum number. The latter processes, which involve absorption of circularly polarized photons (designated $\sigma+$ and $\sigma-$ for helicity $+1$ and -1, respectively[1]) are basically of two kinds. The first, investigated initially by Kusch[2] and by Brossel, Cagnac, and Kastler,[3] entails the absorption of a number p of $\sigma+$ (or $\sigma-$) rf photons by an atom in magnetic substate M with subsequent transition via $p-1$ intermediate atomic states to the final magnetic substate $M+p$ (or $M-p$), as diagrammed in the first frame of figure 4.1. The second, observed initially by Margerie and Brossel,[4] differs from the preceding kind in that the terminal atomic states of the transition also serve as intermediate states (as shown in the second frame of the figure). In both processes, the initial and terminal states have the same parity and lifetime and are coupled to the rf field via the magnetic dipole op-

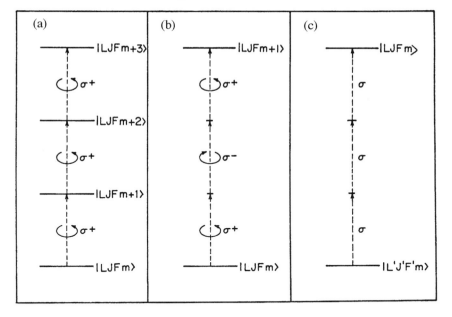

Figure 4.1 Examples of multiple-quantum transitions: (a) triple-quantum magnetic dipole transition with intermediate atomic substates; (b) triple-quantum magnetic dipole transition in which the terminal states also serve as intermediate states; (c) triple-quantum electric dipole transition between hyperfine states in different fine structure levels (subject to selection rules $|\Delta L| = 1$, $|\Delta J| = 1$, 0, $|\Delta F| = 1$, 0).

erator. By contrast, the m.q. transitions of interest here—shown in the third frame—are mediated by the electric dipole operator and involve terminal states of opposite parity; the absorbed photons are linearly polarized, and there is no change in angular momentum along the axis of quantization (defined by the electric field).

In this chapter, we will examine the dynamics of the atom-field interaction from the broader perspective of nonrelativistic quantum electrodynamics (QED), derive equations for the time evolution of the amplitudes of the combined atom–rf field states, and compare the results of the quantized field method with those of the previous classical field approach.[5] The method of solution makes use of a phase transformation introduced in the preceding chapters, the significance of which will be seen shortly as a resetting of the origin of the energy scale.

Let us consider, therefore, a quantum system comprising a multistate atom and a monochromatic linearly polarized rf field. The basis states of such a system can be written as the direct product $|\mu; n\rangle = |\mu\rangle|n\rangle$, where μ labels completely the state of the atom and n is the number of photons present in the field; each photon is characterized by energy $\hbar\omega$, momentum $\mathbf{p} = \hbar\mathbf{k} = \hbar(\omega/c)\hat{\mathbf{k}}$, and polarization $\hat{\mathbf{e}}_\sigma(\mathbf{k})$ with polarization label $\sigma = 1, 2$ for two orthogonal orientations of the electric field vector. (If the rf field were circularly or, more generally, elliptically polarized, one could write $\sigma = \pm 1$ for two orthogonal states of opposite helicity.) In the preceding symbolism, the caret signifies a unit vector—and whenever there is no need to distinguish different polarization states, the unit electric field vector will be represented simply by $\hat{\mathbf{e}}$.

In the absence of any mutual interaction, the dynamics of the system is prescribed by the Hamiltonian

$$\mathcal{H}_0 = \mathcal{H}_a + \mathcal{H}_{\text{rf}}, \tag{1}$$

in which the atomic Hamiltonian (in units of angular frequency) is

$$\mathcal{H}_a = \sum_\mu (\omega_\mu - i\tfrac{1}{2}\gamma_\mu)|\mu\rangle\langle\mu| \tag{2}$$

and the Hamiltonian of the field (with neglect of zero-point energy, which plays no role under the present circumstances) takes the form

$$\mathcal{H}_{\text{rf}} = \omega a_{\mathbf{k}\sigma}^\dagger a_{\mathbf{a}\sigma} \equiv \omega N_{\mathbf{k}\sigma}. \tag{3}$$

Since the field consists of but a single mode—unique frequency, polarization, and direction of propagation—there is no need at this point to encumber the notation for photon creation $(a^\dagger)$ and annihilation (a) operators, or the photon number operator $N = a^\dagger a$. These operators satisfy the commutation relations

$$[a, a^\dagger] = 1, \qquad [a^\dagger, a^\dagger] = [a, a] = 0, \tag{4a}$$

$$[a, N] = a, \qquad [a^\dagger, N] = -a^\dagger, \tag{4b}$$

from which follow their action on photon-number states:

$$a|n\rangle = \sqrt{n}|n - 1\rangle, \tag{5a}$$

$$a^\dagger|n\rangle = \sqrt{n + 1}|n + 1\rangle, \tag{5b}$$

$$N|n\rangle = n|n\rangle. \tag{5c}$$

For a fuller explanation of the algebraic properties of these operators and derivation of the "harmonic oscillator" Hamiltonian (3), the reader can consult almost any introductory quantum mechanics text.[6] If the set of amplitudes $\{c^0_{\mu;n}\}$ defines the atom-field state at the initial time $t_0 = 0$, then the state evolves under the action of $e^{-i\mathcal{H}_0 t}$ to become

$$\Psi(t) = \sum_{\mu;n} c^0_{\mu;n} e^{-i(\omega_\mu + n\omega - \frac{1}{2} i\gamma_\mu)t} |\mu; n\rangle. \tag{6}$$

In treating the interaction between the atom and the field, it serves our purposes best here to begin *not* with the explicit electric dipole interaction $(-\boldsymbol{\mu}_E \cdot \mathbf{E})$ used earlier [Eq. (2.1)], but rather with the Hamiltonian[7]

$$\mathcal{H}_I = -\frac{(-e)}{mc} \mathbf{A} \cdot \mathbf{p}, \tag{7}$$

in which $\mathbf{p}$ is the electron linear momentum operator and

$$\mathbf{A}(\mathbf{r}) = \alpha\left(a\hat{\mathbf{e}}e^{i\mathbf{k}\cdot\mathbf{r}} + a^\dagger\hat{\mathbf{e}}^* e^{-i\mathbf{k}\cdot\mathbf{r}}\right) \tag{8a}$$

is the vector potential of the field in the transverse gauge; that is, the gauge for which $\nabla \cdot \mathbf{A} = 0$. (The relationship between the "$\mathbf{A} \cdot \mathbf{p}$" and "$\boldsymbol{\mu}_E \cdot \mathbf{E}$" interactions is discussed in the next section.) The coefficient α in Eq. (8a) is given by

$$\alpha = \sqrt{\frac{2\pi c^2}{\omega \mathcal{V}}} \tag{8b}$$

with $\mathcal{V}$ the volume within which the field is quantized. [To express $\mathcal{H}$ in energy units, insert $\hbar$ in the numerator of Eq. (8b).] Again, only one electromagnetic mode is presumed present in the field, so that no sum over $\mathbf{k}$, σ is required. With the interaction between field and atom turned on, the time evolution of the state vector

$$\Psi(t) = \sum_{\mu;n} c_{\mu;n}(t) |\mu; n\rangle \tag{9}$$

is governed by the Schrödinger equation

$$i\frac{\partial\Psi(t)}{\partial t} = \mathcal{H}\Psi(t) = (\mathcal{H}_0 + \mathcal{H}\ )\Psi(t). \tag{10}$$

Before considering solutions to Eq. (10), let us first examine the relationship between the quantized vector potential operator, Eq. (8a), and the classical vector potential field

$$\mathbf{A}_{cl} = \mathbf{A}_0 \cos(\omega t + \delta), \tag{11}$$

in which δ is a well-defined relative phase. This relationship derives from the correspondence

$$\sqrt{\langle\mathbf{A}_{cl}\cdot\mathbf{A}_{cl}\rangle_t} \leftrightarrow \sqrt{\langle n|\mathbf{A}(\mathbf{r})\cdot\mathbf{A}(\mathbf{r})|n\rangle} \tag{12}$$

where is a $\langle\ \rangle_t$ time average over a full cycle of the classical field. Evaluating both sides of Eq. (12) leads to

$$\frac{1}{\sqrt{2}}A_0 = \alpha\sqrt{2n+1}, \tag{13a}$$

or

$$\alpha \approx \frac{A_0}{2\sqrt{n}} \tag{13b}$$

for classical fields with $n \gg 1$.

To the above correspondence may be raised an objection that the photon occupation number of a classical field is an indeterminate quantity and that the so-called "coherent" states (first studied by Schrödinger and introduced into quantum optics by Glauber[8]) provide the proper correspondence between classical and quantum electromagnetic fields. These states, constructed from the following linear superposition of occupation number states,

$$|a\rangle = e^{-\frac{1}{2}|a|^2}\sum_{n=0}^{\infty}\frac{a^n}{\sqrt{n!}}|n\rangle, \tag{14}$$

are eigenvectors of the annihilation operator a,

$$a|a\rangle = a|a\rangle, \tag{15a}$$

in which the eigenvalue a is a complex-valued parameter. The conjugate states are eigenvectors of the creation operator $a^\dagger$,

$$\langle a|a^\dagger = \langle a|a^*. \tag{15b}$$

From the defining relation (14) one can show that a is related to the mean photon occupation number $\bar{n}$ according to

$$a = \sqrt{\bar{n}}e^{i\delta}. \tag{16}$$

The amplitude of the classical field is given by the expectation value of $\mathbf{A}(\mathbf{r})$ in a coherent state:

$$\langle a|\mathbf{A}|a\rangle = A_0\mathbf{e}. \tag{17}$$

Equation (16) shows that a definite phase δ can be associated with coherent states, and therefore with classical fields. As discussed in the previous chapter, however, the phase of the rf field in a fast-beam experiment is a distributed quantity; it is different for atoms passing through the field at different points in the oscillation cycle, and we will ultimately average phase-dependent quantities over all values of δ from 0 to 2π. For the present, this is equivalent to taking $a = \sqrt{\bar{n}}$ at the outset. Since the wavelength of a rf photon is much larger than atomic dimensions, that is, $\mathbf{k} \cdot \mathbf{r} \ll 1$, it is permissible to make the dipole approximation $e^{\pm i\mathbf{k}\cdot\mathbf{r}} \sim 1$ in Eq. (8a). Equation (17) then leads to the result

$$\alpha = \frac{A_0}{2\sqrt{\bar{n}}}, \tag{18}$$

which differs from Eq. (13b) only by replacement of the sharp occupation number n by the mean $\bar{n}$. The closeness of these two expressions can be illustrated by determining the dispersion about $\bar{n}$, that is,

$$\Delta n = \sqrt{\langle a|N^2|a\rangle - (\langle a|N|a\rangle)^2} = \sqrt{\bar{n}}. \tag{19}$$

Hence

$$\frac{\Delta n}{\bar{n}} = \frac{1}{\sqrt{\bar{n}}} \ll 1 \tag{20}$$

if $\bar{n} \gg 1$.

Using Eq. (8b) for α and the relationship

$$\mathbf{E}_{cl} = -\frac{\partial \mathbf{A}_{cl}}{c \partial t} \Rightarrow A_0 = \frac{cE_0}{\omega} \tag{21}$$

for the amplitudes of the classical vector potential and electric fields, one obtains the following expression (with $\hbar$ temporarily restored to show dimensional consistency) for the mean number of photons per unit volume:

$$\frac{\bar{n}}{\mathcal{V}} = \frac{E_0^2}{8\pi\hbar\omega}. \tag{22}$$

Equation (22) also follows straightforwardly from the relation for energy density of the electric field: $u_E = \bar{n}\hbar\omega/\mathcal{V} = E_0^2/8\pi$. Consider, for example, a rf field oscillating at 100 MHz with an amplitude $E_0 = 1$ V/cm $= 1/300$ statvolt/cm. Then the preceding equations lead to a photon density $\bar{n}/\mathcal{V} \sim 6.7 \times 10^{11}$ cm^{-3} and relative dispersion $\Delta n/\bar{n} \sim 1.2 \times 10^{-6}$. One sees that the mean photon number of a classical field is enormous and fluctuations negligibly small. Hence we may describe a classical field quite adequately by the occupation state $|n\rangle$ for $n \sim \bar{n}$ in place of the coherent state $|a\rangle$ if we give up all knowledge of field phase.[9]

Returning to Eq. (7), we insert the quantized vector potential operator of Eq. (8a) (for a field linearly polarized along the z-axis) to obtain the interaction Hamiltonian (again in units of angular frequency)

$$\mathcal{H} = \frac{\theta}{\sqrt{\bar{n}}} p_z (a + a^\dagger) \tag{23a}$$

with coupling constant

$$\theta = \frac{eA_0}{2mc}. \tag{23b}$$

Use of this Hamiltonian, together with Hamiltonians (2) and (3) for the uncoupled atom and rf field, in the Schrödinger equation (10) leads to the following set of coupled equations for the amplitudes of the total atom-field system:

$$\frac{dc_{\mu;p}(t)}{dt} = -\left[i(\omega_\mu + p\omega) + \tfrac{1}{2}\gamma_\mu\right]c_{\mu;p}(t)$$

$$-i\sum_\nu V_{\mu\nu}\left[\sqrt{\frac{p}{\bar{n}}}c_{\nu;p-1}(t) + \sqrt{\frac{p+1}{\bar{n}}}c_{\nu;p+1}(t)\right] \quad (24)$$

with matrix elements

$$V_{\mu\nu} = V_{\nu\mu} = \langle\mu|\theta p_z|\nu\rangle. \quad (25)$$

In the infinite set of equations (24), each amplitude $c_{\mu;p}(t)$ is coupled to itself and to amplitudes of states of opposite parity and photon numbers increased or diminished by one. Since $\bar{n} \sim p \sim p \pm 1$ for all processes of interest here, we can replace the radicals in Eq. (24) by unity, while still retaining the distinct photon number labels on all amplitudes.

4.2 REMARKS ON DIPOLE COUPLING

In the semiclassical treatment of the atom–rf field coupling of the previous two chapters, the interaction Hamiltonian took the form $\mathcal{H}' = -\boldsymbol{\mu}_E \cdot \mathbf{E}_0 \cos\omega t$ (with omission of the phase parameter δ which is not relevant to the present discussion). This led to the electric dipole matrix element of Eq. (3.18), here designated $V'_{\mu\nu} = \langle\mu| - \tfrac{1}{2}\boldsymbol{\mu}_E \cdot \mathbf{E}|\nu\rangle$. It is instructive to comment briefly on the relationship between $V'_{\mu\nu}$ and the element $V_{\mu\nu} = \langle\mu|(e/2mc)\mathbf{A}_0 \cdot \mathbf{p}|\nu\rangle$ equivalent to Eq. (25).

We start with the familiar nonrelativistic Hamiltonian for a particle of arbitrary charge q (with $q = -e$ for the electron):

$$\mathcal{H} = \frac{1}{2m}\left(\mathbf{p} - \frac{q\mathbf{A}}{c}\right)^2 + U(\mathbf{r}) = \mathcal{H}_0 - \frac{q}{mc}\mathbf{A}\cdot\mathbf{p} + \frac{q^2}{2mc^2}\mathbf{A}\cdot\mathbf{A} \quad (26)$$

from which $V_{\mu\nu}$ derives. $\mathbf{A}$ is a classical vector potential field (in the transverse gauge) assumed here to be homogeneous (that is, independent of spatial coordinates), and $U(\mathbf{r})$ is a time-independent potential term. If Ψ is a solution of the Schrödinger equation (10), then an equally valid solution leading to the same physical predictions is obtained from the unitary transformation

$$\Psi = e^{-iq\mathbf{A}\cdot\mathbf{r}/c}\Psi'. \tag{27}$$

Substitution of Eq. (27) into Eq. (10) leads to an equation of the same form if, in place of $\mathcal{H}$, we have

$$\mathcal{H}' = e^{-iq\mathbf{A}\cdot\mathbf{r}/c}\frac{\left(\mathbf{p} - \dfrac{q}{c}\mathbf{A}\right)^2}{2m}e^{iq\mathbf{A}\cdot\mathbf{r}/c} + \frac{q}{c}\frac{\partial\mathbf{A}}{\partial t}\cdot\mathbf{r} + U(\mathbf{r}). \tag{28}$$

Careful evaluation of the operator $\mathcal{H}'$ and use of Eq. (21) defining the electric field $\mathbf{E}$ leads to the expression

$$\mathcal{H}' = \mathcal{H}_0 - \boldsymbol{\mu}_E \cdot \mathbf{E} \tag{29}$$

with dipole operator $\boldsymbol{\mu}_E = q\mathbf{r}$.

It is seen therefore that the electric dipole interaction term in the Hamiltonian (29) corresponds to terms $-(q/mc)\mathbf{A}\cdot\mathbf{p} + (q^2/2mc^2)\mathbf{A}\cdot\mathbf{A}$ in the Hamiltonian (26). Hence the matrix elements $V_{\mu\nu}$ and $V'_{\mu\nu}$ should be equivalent since the "diamagnetic" term quadratic in the vector potential cannot couple two orthogonal atomic states. Similar results were first obtained many years ago by Maria Göppert-Mayer,[10] who took as her starting point the classical Lagrangian operator $L(\mathbf{r}, \dot{\mathbf{r}}) = \frac{1}{2}m\dot{\mathbf{r}}^2 - U(\mathbf{r}) - (q/c)\dot{\mathbf{r}}\cdot\mathbf{A}$ (with $\dot{\mathbf{r}} \equiv d\mathbf{r}/dt$) and transformed it by adding the total time derivative $(d/dt)(\mathbf{r}\cdot\mathbf{A})$.

It has long been known,[11] however, that a sort of paradox arises when one begins with the standard operator equation

$$\mathbf{p} = m\dot{\mathbf{r}} = -im[\mathbf{r}, \mathcal{H}_0] \tag{30}$$

and uses Eq. (21) to relate the amplitudes of the classical electric and vector potential fields. Substitution into $V_{\mu\nu}$ of Eq. (30) for $\mathbf{p}$ and $A_0 = cE_0/\omega$ yields a relation

$$V_{\mu\nu} = i\frac{\omega_{\mu\nu}}{\omega}V'_{\mu\nu} \tag{31}$$

with the erroneous implication of a low-frequency divergence. This result is not correct, but discussion of the error and its rectification goes beyond the scope of this book. Suffice it to say that, when used consistently, the "$\mu_E \cdot \mathbf{E}$" and "$\mathbf{A} \cdot \mathbf{p}$" forms of the dipole interaction lead to equivalent results. For a comprehensive treatment of the problem, the reader should consult the reference by Power and Zienau.[12]

4.3 THE TWO-LEVEL ATOM (AGAIN)

We restrict the subsystem of atomic states to two, $\mu = \alpha$, β with $\omega_\alpha > \omega_\beta$, and substitute into Eq. (24) the amplitudes

$$c_{\alpha;p}(t) = a_{\alpha;p}(t)e^{-i(\omega_\alpha + p\omega)t}, \tag{32a}$$

$$c_{\beta;q}(t) = a_{\beta;q}(t)e^{-i(\omega_\beta + q\omega)t}, \tag{32b}$$

to obtain the following coupled equations in the interaction representation:

$$\frac{da_{\alpha;p}(t)}{dt} = -\tfrac{1}{2}\gamma_\mu a_{\alpha;p}(t) - iV\left[a_{\beta;p-1}(t)e^{i\Omega't} + a_{\beta;p+1}(t)e^{-i\Omega t}\right], \tag{33a}$$

$$\frac{da_{\beta;q}(t)}{dt} = -\tfrac{1}{2}\gamma_\beta a_{\beta;q}(t) - iV\left[a_{\alpha;q-1}(t)e^{i\Omega t} + a_{\alpha;q+1}(t)e^{-i\Omega't}\right], \tag{33b}$$

with matrix element $V = V_{\alpha\beta}$, detuning frequency $\Omega = \omega - \omega_0$, antiresonance frequency $\Omega' = \omega + \omega_0$, and Bohr frequency $\omega_0 = \omega_\alpha - \omega_\beta$, in accordance with the notation of earlier chapters.

If we set $q = p + 1$, Eqs. (33a,b) are seen to resemble Eq. (2.11), which was derived for a classical rf field coupling two atomic states. The amplitude $\langle 1|\Psi \rangle$ in (2.11) replaces $a_{\alpha;p}$ and $a_{\alpha;p+2}$ above, while amplitude $\langle 2|\Psi \rangle$ in Eq. (2.11) replaces $a_{\beta;p+1}$ and $a_{\beta;p-1}$. Actually, it will be shown later that the amplitudes $\langle 1|\Psi \rangle$ and $\langle 2|\Psi \rangle$ are respectively related to $a_{\alpha;p+k}$ (k even) and $a_{\beta;p+k}$ (k odd) multiplied by a phase factor and summed over all values of k from $-p$ (effectively from $-\infty$ since $p \gg 1$) to $+\infty$. The resulting transition probability therefore encompasses all allowed multiple-quantum transition probabilities plus contributions from what could be termed interference effects between multiphoton processes of different order. In many experimental situations these interference terms are averaged to zero. Within the framework of the usual semiclassical treatment, the existence of multiquantum processes does not emerge clearly, nor can one determine the properties of the resonance lineshape for transitions of a given order. These points will be taken up in more detail shortly.

The purpose of deriving Eqs. (33a, b) was primarily to point out the similarity between the semiclassical equations of motion obtained in chapter 2 and those derived by the quantized field formalism. However, transforming Eq. (24) into (33a, b) is not the best way to proceed to solve the equations. Indeed, it detracts from one of the main advantages of the quantized field approach, namely, to obtain at the outset a set of equations whose only time dependence is intrinsic to the unknown amplitudes and not contained additionally in exponential phase factors. To resolve this problem we make use of a phase transformation employed in earlier chapters to reset the zero of energy in such a manner as to produce time-independent coefficients of comparable order of magnitude.

Start again with Eq. (1) and rewrite the Hamiltonian $\mathcal{H}_0$ as

$$\mathcal{H}_0 = W - \tfrac{1}{2} i \Gamma_D, \tag{34a}$$

where the energy (actually frequency) operator

$$W = \sum_{\mu;n} (\omega_\nu + n\omega) |\mu; n\rangle \langle \mu; n| \equiv \sum_{\mu;n} W_{\mu;n} |\mu; n\rangle \langle \mu; n| \tag{34b}$$

generates the eigenvalues of the noninteracting atom-field system, and the decay operator

$$\Gamma_D = \sum_{\mu;n} \gamma_\mu |\mu; n\rangle \langle \mu; n| \tag{34c}$$

accounts phenomenologically for the finite lifetime of the states.[13] The state vector $\Psi(t)$ is now transformed to the vector $\Psi_\Lambda(t)$,

$$\Psi(t) = e^{-i(W+\Lambda)t} \Psi_\Lambda(t), \tag{35}$$

in which Λ is a diagonal operator in the space spanned by the compound kets $\{|\mu; n\rangle\}$,

$$\Lambda |\mu; n\rangle = \lambda_{\mu;n} |\mu; n\rangle. \tag{36}$$

Substitution of Eqs. (34a–c) through (36) into the Schrödinger equation then yields the set of equations

$$\frac{d}{dt}(\Psi_\Lambda)_{\mu;\,p} = \sum_{\nu;\,q} \Pi_{\mu;\,p\,|\,\nu;\,q}(\Psi_\Lambda)_{\nu;\,q},\tag{37}$$

in which

$$(\Psi_\Lambda)_{\mu;\,p} = \langle \mu;\,p \mid \Psi_\Lambda \rangle\tag{38}$$

and

$$\Pi_{\mu;\,p\,|\,\nu;\,q} = -i\big[\langle \mu;\,p|\mathcal{H}\,|\nu;\,q\rangle e^{i(W_{\mu;\,p}-W_{\nu;\,q}+\lambda_{\mu;\,p}-\lambda_{\nu;\,q})t}$$
$$-\,(\lambda_{\nu;\,q} + \tfrac{1}{2}i\gamma_\nu)\delta_{\mu\nu}\delta_{pq}\big].\tag{39}$$

Our task is to seek the eigenvalues of Λ such that the matrix Π is independent of time. This procedure was followed in chapters 2 and 3, but no specific explanation was then given for the choice of solutions.

It is apparent from Eq. (39) that the eigenvalue $\lambda_{\mu;\,p}$ can be chosen up to a constant κ to be

$$\lambda_{\mu;\,p} = -W_{\mu;\,p} + \kappa = -(\omega_\mu + p\omega) + \kappa.\tag{40a}$$

The constant κ allows one to reset the zero of energy and, by so doing, to transform the equations into the interaction representation or any of the generalized rotating coordinate frames. We take as our energy level standard a particular value $\kappa = \omega_\sigma + s\omega$ from which follows the eigenvalue

$$\lambda_{\mu;\,p} = -\omega_{\mu\sigma} - (p - s)\omega\tag{40b}$$

with $\omega_{\mu\sigma} = \omega_\mu - \omega_\sigma$. The matrix Π then assumes the time-independent form

$$\Pi_{\mu;\,p\,|\,\nu;\,q} = -i\big[\langle \mu;\,p|\mathcal{H}\,|\nu;\,q\rangle$$
$$+\,(\omega_{\nu\sigma} + (q - s)\omega - \tfrac{1}{2}i\gamma_\nu)\delta_{\mu\nu}\delta_{pq}\big].\tag{41}$$

Equation (37), with the Hamiltonian matrix Π defined by Eq. (41), yields the set of equations we require. Since the set is infinite, one must in practice truncate the set by including as many amplitudes as are necessary to calculate the probability of a m.q. transition of designated order. We will implement this procedure in a specific case shortly, but for the present let us continue to maintain complete generality. The formal solution to Eq. (37) is obtained by first transforming the time-independent matrix Π to its diagonal form ε

$$\Pi = C\varepsilon C^{-1}, \tag{42}$$

by means of the matrix C whose columns are the eigenvectors of Π (as discussed in section 2.2). The resulting equation can then be integrated immediately to give

$$\Psi_\Lambda(t) = \left[Ce^{\varepsilon\tau}C^{-1} \right]\Psi_\Lambda(t_0) \equiv M(\tau)\Psi_\Lambda(t_0) \tag{43}$$

where the duration of the interaction is $\tau = t - t_0$. It then follows from Eq. (35) that

$$\Psi_\Lambda(t) = e^{-i(W+\Lambda)t}M(\tau)e^{i(W+\Lambda)t_0}\Psi_\Lambda(t_0) \tag{44}$$

is the complete solution to the Schrödinger equation for the combined atom-field system.

As an example, let the atom be in state $|\alpha\rangle$ and the field contain p rf photons

$$\Psi(t_0) = |\alpha; p\rangle \tag{45}$$

at the initial time of their interaction, t_0. The probability for transition to the atomic state $|\beta\rangle$ with k photons participating is then

$$P_{\beta;p+k\,|\,\alpha;\,p}(\tau) = \left| \langle \beta; p+k | e^{-i(W+\Lambda)t}M(\tau)e^{i(W+\Lambda)t_0} |\alpha; p\rangle \right|^2$$
$$\equiv \left| M_{\beta;\,p+k\,|\,\alpha;\,p}(\tau) \right|^2 \tag{46}$$

and depends only on the duration of the interaction, not on the initial time. To calculate the total probability of an atomic transition irrespective of the number of absorbed or emitted photons, Eq. (46) must be summed over all k:

$$P_{\beta|\alpha}(\tau) = \sum_{k=-p}^{\infty} \left| M_{\beta;p+k\,|\,\alpha;\,p}(\tau) \right|^2. \tag{47}$$

For a classical field the number of photons p is very large and to good approximation the lower limit of the sum can be replaced by $k = -\infty$.

4.4 COHERENT FIELD STATES

In a classical rf field the number of photons is not known with absolute certainty. That the artificiality of the intial condition, Eq. (45), does not lead to neglect of interference terms in the total transition probability, Eq. (47), or to any other loss of generality can be demonstrated by use of the coherent field states. To see this, we choose the initial state of the system to be

$$\Psi(t_0) = |\alpha; a\rangle, \tag{48}$$

where, from Eqs. (14) and (16), the coherent state $|a\rangle$ can be written as

$$|a\rangle = \sum_p \chi_p(\delta)|p\rangle \tag{49a}$$

with expansion coefficients

$$\chi_p(\delta) = e^{-\frac{1}{2}|a|^2} \frac{|a|^p}{\sqrt{p!}} e^{ip\delta}. \tag{49b}$$

Recall that $|a|^2$ is the mean photon occupation number $\bar{n}$, and $\delta = -\omega t_0$ sets the initial phase of the field.

The amplitude of a transition to the final state $|\beta; a\rangle$—note that the classical field remains a coherent state—is then given by the expression

$$e_{\beta;a}(t) = \sum_{(p,q)=0}^{\infty} \chi_q^*(\delta)\chi_p(\delta)\langle\beta; q|e^{-i(W+\Lambda)t}M(\tau)e^{i(W+\Lambda)t_0}|\alpha; p\rangle, \tag{50}$$

which, upon rearranging phase factors and setting the summation index $q = p + k$, leads to

$$c_{\beta;a}(t) = \sum_{p=0}^{\infty} \sum_{k=-p}^{\infty} \chi_{p+k}^*(\delta)\chi_p(\delta)$$
$$\times e^{i(W_{\alpha;p}-W_{\beta;p+k}+\lambda_{\alpha;p}-\lambda_{\beta;p+k})t_0} e^{-i(W_{\beta;p+k}+\lambda_{\beta;p+k})\tau}$$
$$\times M_{\beta;p+k\,|\,\alpha;p}(\tau). \tag{51}$$

To simplify Eq. (51), recall again that for a classical field the deviation in photon number from the mean value $\bar{n}$ is negligibly small, so that there is effectively only one contribution ($p \sim \bar{n}$) to the sum over p, and the lower range of k can be extended to $-\infty$ with no observable consequences. With p restricted to $\bar{n}$ and the standard energy scale defined by $\omega_\sigma \equiv \omega_\beta$ and $s \equiv p \approx \bar{n}$, we obtain the relations

$$\lambda_{\alpha; p} = -\omega_{\alpha\beta} = -\omega_0, \tag{52a}$$

$$\lambda_{\beta; p+k} = -k\omega. \tag{52b}$$

Upon substitution of the elements (52a,b) into Eq. (51), the first exponential phase factor is eliminated [thereby leaving only the field coefficients $\chi(\delta)$ dependent on initial conditions], and the second exponential phase factor becomes $e^{-i(\omega_\beta + p\omega)\tau}$, independent of the index k. The expression for the total atomic transition probability then becomes

$$P_{\beta|\alpha}(t) = \left| \sum_{k=-\infty}^{\infty} \chi_{p+k}^*(\delta)\chi_p(\delta)M_{\beta; p+k|\alpha; p}(\tau) \right|^2. \tag{53}$$

From Eq. (49b) the product of field amplitudes in Eq. (53) is given by

$$\chi_{p+k}^*(\delta)\chi_p(\delta) = \frac{e^{|a|^2}|a|^{2p+k}}{\sqrt{(p+k)!\,p!}}e^{ik\delta} \rightarrow \frac{e^{-\bar{n}}\bar{n}^{\bar{n}}}{\bar{n}!}e^{ik\delta} \tag{54a}$$

and reduces to a simpler expression, as indicated by the arrow, under the circumstance $p \gg k$, valid for a classical field. Application of Stirling's approximation ($\ln x! \approx x \ln x - x$ for $x \gg 1$), further simplifies Eq. (54a) to

$$\chi_{p+k}^*(\delta)\chi_p(\delta) \approx e^{ik\delta}, \tag{54b}$$

with the result that the atomic transition probability

$$P_{\beta|\alpha}(t) = \left| \sum_{k=-\infty}^{\infty} e^{ik\delta}M_{\beta; p+k|\alpha; p}(\tau) \right|^2, \tag{55}$$

when averaged over the range 0 to 2π of the field phase, is precisely the transition probability of Eq. (47) derived for the initial state $|\alpha; p\rangle$. Thus we see again that a single occupation number state is an adequate representation of the classical rf field when no specific phase information is utilized.

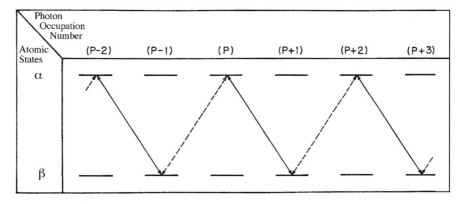

Figure 4.2 Triple-quantum process treated in a six-state approximation. The initial state of the system is $|\alpha; p\rangle$ and the final state is $|\beta; p + 3\rangle$. Real transitions are indicated by solid arrows and virtual transitions by dashed arrows.

4.5 TRIPLE-QUANTUM TRANSITIONS

As an application of the preceding theory, we will now examine the lowest-order $(r > 1)$ multiphoton process that can be induced between two hydrogen fine structure states by a rf electric field parallel to the axis of quantization. It is assumed that at time t_0 the atom-field system is in the state $|\alpha; p\rangle$ where $|\alpha\rangle$ is an eigenfunction of the field-free hydrogen Hamiltonian. Inspection of figure 4.2 shows that a stimulated three-photon transition that brings the system into the state $|\beta; p + 3\rangle$ is the lowest allowed multiple-quantum process; there is no real double-quantum transition between atomic states of opposite parity.

To treat this triple-quantum process we retain in the state vector the six time-dependent amplitudes $\{c_{\alpha;\,p-2},\ c_{\alpha;\,p-1},\ c_{\alpha;\,p},\ c_{\alpha;\,p+1},\ c_{\alpha;\,p+2},\ c_{\alpha;\,p+3}\}$ and employ the standard energy scale leading to the matrix Π of Eq. (41), which reduces in this case to

$$
\Pi = \begin{pmatrix}
-i(\omega_0 - 2\omega - i\tfrac{1}{2}\gamma_\alpha) & -iV & 0 & 0 & 0 & 0 \\
-iV & -i(\omega - i\tfrac{1}{2}\gamma_\beta) & -iV & 0 & 0 & 0 \\
0 & -iV & -i(\omega_0 - i\tfrac{1}{2}\gamma_\alpha) & -iV & 0 & 0 \\
0 & 0 & -iV & -i(\omega - i\tfrac{1}{2}\gamma_\beta) & -iV & 0 \\
0 & 0 & 0 & -iV & -i(\omega_0 + 2\omega - i\tfrac{1}{2}\gamma_\alpha) & -iV \\
0 & 0 & 0 & 0 & -iV & -i(3\omega - i\tfrac{1}{2}\gamma_\beta)
\end{pmatrix}.
$$

$$(56)$$

The diagonal element Π_{33} is the (complex) energy eigenvalue of the initial state $|\alpha; p\rangle$, in accordance with the above ordering of state amplitudes. Application of standard matrix algebra then leads to the transition matrix $M(\tau)$, defined by Eq. (43), from which the resonance lineshapes for single- and triple-quantum transitions are obtained by evaluating

$$P_{\beta;\, p+1\,|\,\alpha;\, p}(\tau) = \left|M_{\beta;\, p+1\,|\,\alpha;\, p}(\tau)\right|^2 = \left|M_{43}(\tau)\right|^2, \tag{57a}$$

$$P_{\beta;\, p+3\,|\,\alpha;\, p}(\tau) = \left|M_{\beta;\, p+3\,|\,\alpha;\, p}(\tau)\right|^2 = \left|M_{63}(\tau)\right|^2, \tag{57b}$$

as a function of rf frequency ω; numerical subscripts (corresponding to those of Π) locate the element in the 6×6 matrix. Within the six-state approximation the total probability of a transition $|\alpha\rangle \rightarrow |\beta\rangle$ is

$$P_{\beta\,|\,\alpha}(\tau) = \left|M_{\beta;\, p-1\,|\,\alpha;\, p}(\tau)\right|^2 + \left|M_{\beta;\, p+1\,|\,\alpha;\, p}(\tau)\right|^2 + \left|M_{\beta;\, p+3\,|\,\alpha;\, p}(\tau)\right|^2$$

$$= \sum_{k=1}^{3} \left|M_{2k,\, 3}(\tau)\right|^2 \tag{58a}$$

and correspondingly

$$P_{\alpha;\, p}(\tau) = \left|M_{\alpha;\, p\,|\,\alpha;\, p}(\tau)\right|^2 = \left|M_{33}(\tau)\right|^2 \tag{58b}$$

gives the occupation probability of the initial state.

Although the explicit analytical expressions to which Eqs. (57a, b) and (58a, b) give rise are useful for numerical calculation and graphical display of lineshapes, they are algebraically complicated and do not provide much physical insight. A deeper understanding of the physics of multiphoton processes can be obtained by an alternative perspective described in the following section.

4.6 CROSSINGS AND ANTICROSSINGS

A significant advantage of the quantized field approach, apart from the fact that the interaction Hamiltonian is independent of time, is the insight to be gained from an examination of the energy level structure of the combined atom-field system.

Consider, for example, figure 4.3, which shows the variation in system energy levels as a function of rf frequency ω in the absence of

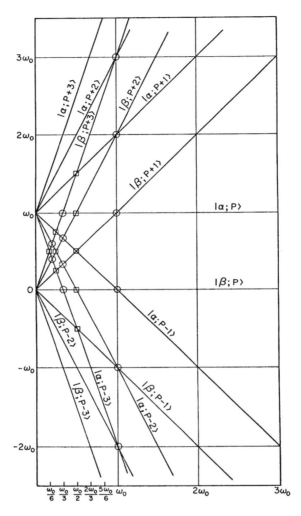

Figure 4.3 Energy level diagram of the noninteracting atom–rf field system. When the interaction is "turned on," level intersections marked by a square become crossing points while those marked by a circle become anticrossing points.

an interaction (mediated by $\mathcal{H}_I$) between the atom and the field. We assume that a phase transformation has been made so that the zero of the energy scale is set at $W_{\beta;\,p}$. The energy $W_{\alpha;\,p}$ is then a constant $\omega_0 = \omega_\alpha - \omega_\beta$ above $W_{\beta;\,p}$ independent of the frequency of the rf field. The other energy levels $W_{\mu;\,q}$ ($\mu = \alpha, \beta; q \neq p$) vary linearly in ω with slope $(\partial/\partial\omega)(W_{\mu;\,q} - W_{\beta;\,p}) = q - p$. Points of level intersection occurring at frequencies $\omega = \omega_0/2n$ (with $n = 1, 2, 3, \ldots$) are marked in the figure with a square; those occurring at frequencies $\omega = \omega_0/(2n+1)$ are marked by a circle. The distinguishing factor is the behavior of the levels at these points upon restoration of the atom-field interaction. Levels intersecting at the squares remain uncoupled to all orders of the interaction Hamiltonian. By contrast, levels intersecting at circles are coupled by $\mathcal{H}_I$ directly or indirectly and, in fact, never actually cross if $V^2 > (\gamma_\alpha - \gamma_\beta)^2/16$. These locations are designated anticrossings[14] and represent an admixture of the unperturbed states.

Figure 4.4 shows the energy level diagram after an interaction of strength $V^2 > (\gamma_\alpha - \gamma_\beta)^2/16$ has been turned on. The dotted lines mark the unperturbed level positions of figure 4.3, while the solid lines, which "repel" each other at the anticrossings, map out the eigenvalues of the total hamiltonian $\mathcal{H} = \mathcal{H}_0 + \mathcal{H}_I$. Anticrossing points in the quantized field description correspond to the locations of m.q. transitions in the language of time-dependent perturbation theory. With the number of levels included in the diagram one can account for single-, triple-, and quintuple-quantum processes.

Before examining aspects of level anticrossings in mathematical detail, it is useful first to make some qualitative observations. In the vicinity of an anticrossing, the separation between repelling levels increases with the coupling strength V. As the coupling diminishes, the levels approach one another, becoming more and more cusped until the coupling strength satisfies the relation $V^2 = (\gamma_\alpha - \gamma_\beta)^2/16$, whereupon the cusped levels touch. For yet weaker couplings, $V^2 < (\gamma_\alpha - \gamma_\beta)^2/16$, the levels do cross. This behavior is illustrated in part a of figure 4.5. Actually, the foregoing description is strictly true only for an isolated pair of coupled states. Should these states be coupled to other states via virtual (that is, energy nonconserving) transitions, the anticrossing point is shifted upward by an amount dependent upon the coupling strength and the unperturbed level separation, as in part b of the figure. Field strengths required for

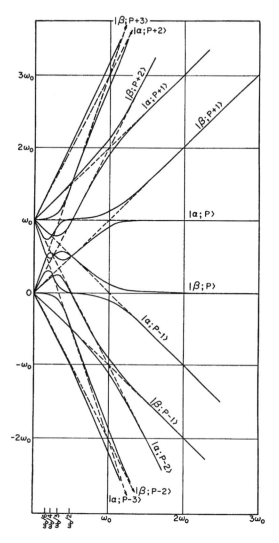

Figure 4.4 Energy level diagram of the interacting atom–rf field system. Dashed lines show the unperturbed levels of figure 4.3. Multiple-quantum transitions correspond to the superposition of unperturbed states in the region of the anti-crossing points.

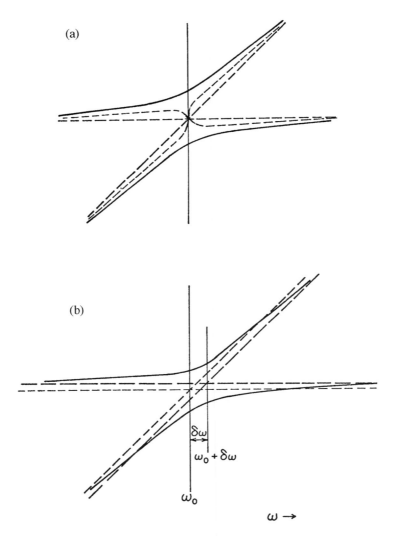

Figure 4.5 Characteristics of the anticrossing region: (a) An isolated pair of levels coupled by an interaction matrix element repel (solid lines) for $V^2 > (\gamma_\alpha - \gamma_\beta)^2/16$ and cross (dashed lines) for $V^2 < (\gamma_\alpha - \gamma_\beta)^2/16$; (b) for a nonisolated pair of coupled levels, the anticrossing point is shifted upward in frequency by $\delta\omega$ and the level decay rates are modified.

observation of the usual single-quantum transitions between hydrogen fine structure states generally give rise to negligible shifts. However, for the more intense fields required to drive multiple-quantum transitions, the shifts can become significant. It is worth noting that shifts can also occur at the level crossing points. In figure 4.4, for example, the crossings at $\omega = \frac{1}{2}\omega_0$ and $\omega = \frac{1}{4}\omega_0$ will approach each other and eventually merge for large enough coupling strength.

A primary lesson to be gained from the preceding discussion and examination of the structure of figure 4.4 is that, in the determination of nonoverlapping multiple-quantum resonance lineshapes, we need consider only the initial and terminal levels of interest in the region of their anticrossing. For the three-photon transition discussed earlier, this means a reduction from six to two in the number of state amplitudes to be evaluated. Let us now examine this suggestion quantitatively.

4.7 RESOLVENT OPERATOR SOLUTION

The state vector characterizing the total atom-field system at some time $t > t_0$ can be expressed in the form of a contour integral[15]

$$\Psi(t) = \frac{1}{2\pi i} \oint_C dE\, e^{-iE(t-t_0)} G(E) \Psi^0 \tag{59}$$

in which

$$G(E) = (E - \mathcal{H})^{-1} = (E - \mathcal{H}_0 - H)^{-1} \tag{60}$$

is known as the resolvent operator, and Ψ^0 is the initial state vector at t_0 (which, with no loss of generality here, will be set to 0). Justification of Eq. (59) and a brief account of the connection between resolvents and propagators are given in appendix 4B. C, a contour extending along the real axis from $-\infty$ to $+\infty$, must be closed in such a way as to satisfy the principle of causality whereby there can be no response of a system prior to a stimulation. A suitable contour is illustrated in figure 4.6. For times $t > 0$, the semicircular portion must lie in the lower half plane (contour C_1) where the imaginary part of the complex integration variable E is negative. In that way the contribution to Eq. (59) from integration along the semicircle vanishes exponentially with radius,

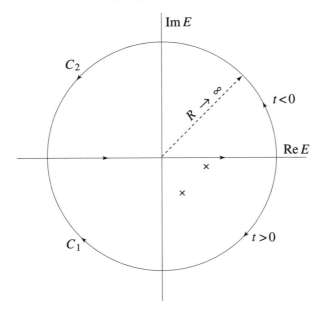

Figure 4.6 Contour on which the integral transforms are to be evaluated. Contour C_1, enclosing two poles with negative imaginary parts, satisfies causality for times $t > 0$.

and only the integration along the real axis remains. This contribution, however, is readily obtained by the Cauchy residue theorem from the poles circumscribed by the entire closed contour.[16] Correspondingly, for times $t < 0$, the semicircular portion must lie in the upper half plane (contour C_2); the state vector then vanishes identically, as causality requires, since the closed contour encompasses no poles and the sum of the residues is zero.

The problem of calculating the state vector $\Psi(t)$ is now reduced to determining $G(E)$, which is usually done by means of the infinite series expansion

$$G(E) = G_0(E) + G_0(E)\mathcal{H}_I G_0(E) + G_0(E)\mathcal{H}_I G_0(E)\mathcal{H}_I G_0(E) + \cdots \tag{61}$$

with

$$G_0(E) = (E - \mathcal{H}_0)^{-1} \tag{62}$$

the resolvent operator for the noninteracting atom-field system. Equations (61) and (62) represent one form of standard perturbation theory.

There is an alternative approach, however, which is particularly appropriate for a region of the energy diagram where only a few states are closely coupled. The resolvent $G(E)$, projected onto a subspace of closely coupled states $\{|\mu; p\rangle\}$, can be expressed as

$$\mathcal{G}(E) \equiv PG(E)P = (E - P\mathcal{H}_0 P - PRP)^{-1}, \tag{63}$$

in which P is the projection operator

$$P = \sum_{\mu; p} |\mu; p\rangle \langle u; p|, \tag{64}$$

Q is the complementary projection operator

$$Q = 1 - P \tag{65}$$

(where **1** is the unit operator), and

$$\mathcal{R} \equiv PRP = P\mathcal{H} \, P + P\mathcal{H} \, P(E - P\mathcal{H}_0 P - Q\mathcal{H} \, Q)^{-1}\mathcal{H} \, P \tag{66}$$

is the residual interaction operator within the subspace of selected states.[17] To implement this approach, we expand the right hand side of Eq. (66) in the infinite series

$$\mathcal{R} = P\big(R^{(r;1)} + R^{(r;2)} + R^{(r;3)} + \cdots\big)P$$
$$= P\big(\mathcal{H} \, + \mathcal{H} \, QG_0\mathcal{H} \, + \mathcal{H} \, QG_0\mathcal{H} \, QG_0\mathcal{H}_i + \cdots\big)P \tag{67}$$

and, as in the case of standard perturbation theory [Eq. (61)] truncate the series at the appropriate order. The superscript r labeling each term in Eq. (67) designates an r-photon quantum process.

As an example of a problem that can be solved exactly, let us reconsider from the viewpoint of the resolvent operator and quantized field formalism the single-quantum coupling of an isolated pair of states. This was treated previously in chapter 2 as the RWA theory. With the initial condition again given by Eq. (45), the state vector evolves in time to become

$$\Psi(t) = \frac{1}{2\pi i} \oint_C dE \, e^{-iEt} \left[G(E)_{\beta; \, p+1 \,|\, \alpha; \, p} |\beta; p+1\rangle + G(E)_{\alpha; \, p \,|\, \alpha; \, p} |\alpha; p\rangle \right],$$

(68)

in which

$$G(E)_{\nu; \, q \,|\, \mu; \, p} = \langle \nu; q | G(E) | \mu; p \rangle.$$

(69)

To calculate the two required elements of $G(E)$ project the exact relation

$$G(E) = G_0(E) + G_0(E) \mathcal{H}_I G(E)$$

(70)

between the states $|\alpha; p\rangle$ and $|\beta; p + 1\rangle$ and solve the resulting two algebraic equations to obtain

$$G(E)_{\beta; \, p+1 \,|\, \alpha; \, p} = \frac{G_0(E)_{\beta; \, p+1 \,|\, \beta; \, p+1} V G_0(E)_{\alpha; \, p \,|\, \alpha; \, p}}{1 - G_0(E)_{\beta; \, p+1 \,|\, \beta; \, p+1} V G_0(E)_{\alpha; \, p \,|\, \alpha; \, p} V}, \quad \text{(71a)}$$

and

$$G(E)_{\alpha; \, p \,|\, \alpha; \, p} = \frac{G_0(E)_{\alpha; \, p \,|\, \alpha; \, p}}{1 - G_0(E)_{\beta; \, p+1 \,|\, \beta; \, p+1} V G_0(E)_{\alpha; \, p \,|\, \alpha; \, p} V}. \quad \text{(71b)}$$

Recall that V is an abbreviated symbol for the element $V_{\alpha\beta}$. From the defining relation (62), the unperturbed resolvent $G_0(E)$ is readily seen to have matrix elements

$$G_0(E)_{\nu; \, q \,|\, \mu; \, p} = \left(E - E^0_{\mu; \, p} - \tfrac{1}{2} i \gamma_\mu \right)^{-1} \delta_{\mu\nu} \delta_{pq}.$$

(72)

After rationalizing the fractions in Eqs. (71a,b), we obtain the expressions

$$G(E)_{\beta; \, p+1 \,|\, \alpha; \, p} = \frac{V}{(E - E^{(+)})(E - E^-)},$$

(73a)

$$G(E)_{\alpha; \, p \,|\, \alpha; \, p} = \frac{E - E_{\beta; \, p+1}}{(E - E^{(+)})(E - E^-)},$$

(73b)

in which

$$E^{(\pm)} = \tfrac{1}{2}\left(E_{\alpha; \, p} + E_{\beta; \, p+1} \right) \pm \tfrac{1}{2} \sqrt{\left(E_{\alpha; \, p} - E_{\beta; \, p+1} \right)^2 + 4V^2} \quad \text{(74)}$$

are the energy eigenvalues of the coupled atom-field system. Thus, the separation or extent of "repulsion" of the two levels is

$$\Delta E = E^{(+)} - E^{(-)} = \sqrt{\left[\omega_0 - \omega - \tfrac{1}{2}i(\gamma_\alpha - \gamma_\beta)\right]^2 + 4V^2} \quad (75a)$$

which, at the anticrossing point $\omega = \omega_0$, reduces to

$$\Delta E = \sqrt{4V^2 - \tfrac{1}{2}(\gamma_\alpha - \gamma_\beta)^2}. \quad (75b)$$

From the above expressions one can demonstrate rigorously the assertions made in the previous section regarding the properties of an isolated pair of coupled states.

When the expressions for $G(E)_{\beta;\,p+1\,|\,\alpha;\,p}$ and $G(E)_{\alpha;\,p\,|\,\alpha;\,p}$ are substituted into the integral of Eq. (59) and evaluated over the contour of figure 4.6, which includes both poles for $t > 0$, the resulting wave amplitudes $c_{\beta;\,p+1}(t)$ and $c_{\alpha;\,p}(t)$ are found to be nearly identical to the amplitudes $c_2(t)$ and $c_1(t)$ of semiclassical RWA theory expressed in Eqs. (2.23) and (2.24a–d). The only difference is that $c_{\beta;\,p+1}(t)$ and $c_{\alpha;\,p}(t)$ are respectively multiplied by the additional phase factors $e^{-i(p+1)\omega t}$ and $e^{-ip\omega t}$ which account for the energy contribution of the field. Inspection of the form of the amplitudes shows immediately that there are no level shifts.

Let us proceed now to treat the general case of any two states $|1\rangle = |\alpha;\,p\rangle$ and $|2\rangle = |\beta;\,q\rangle$ which are coupled either directly or indirectly to some order in $\mathcal{H}$ and thus give rise to an anticrossing. It is now assumed that an infinite number of other levels are present. The simple procedure just employed is inapplicable, and one must resort instead to use of the interaction operator $\mathcal{R} = PRP$ with projection $P = |1\rangle\langle 1| + |2\rangle\langle 2|$. Evaluating the resolvent in the restricted two-dimensional subspace leads to the following matrix representation:

$$\mathcal{G}(E) = (E - \mathcal{H}_0 - \mathcal{R})^{-1}$$

$$= \begin{pmatrix} E - E_1^0 - \mathcal{R}_{11} & -\mathcal{R}_{12} \\ -\mathcal{R}_{21} & E - E_2^0 - R_{22} \end{pmatrix}^{-1}$$

$$= \frac{\begin{pmatrix} E - E_2^0 - \mathcal{R}_{22} & \mathcal{R}_{12} \\ \mathcal{R}_{21} & E - E_1^0 - \mathcal{R}_{11} \end{pmatrix}}{(E - E_2^0 - \mathcal{R}_{22})(E - E_1^0 - \mathcal{R}_{11}) - \mathcal{R}_{12}\mathcal{R}_{21}}. \quad (76)$$

From the form of Eq. (76) it is apparent that the diagonal elements of $\mathcal{R}$ correspond to energy level shifts. Actually, it is the real part of $\mathcal{R}_{11}$ and $\mathcal{R}_{22}$ that produces the energy shifts; the imaginary part leads to a field-dependent shift in state decay rates.

For treatment of a single-photon transition to second order in $\mathcal{H}_I$, we set $q = p + 1$ and evaluate the diagonal elements

$$\mathcal{R}_{kk}^{(1;1)} + \mathcal{R}_{kk}^{(1;2)} = \langle k|\mathcal{H}_I|k\rangle + \langle k|\mathcal{H}_I \frac{Q}{E - \mathcal{H}_0}\mathcal{H}_I|k\rangle \qquad (77)$$

with $k = 1, 2$. By parity consideration (confirmable by explicit calculation), the first-order elements in Eq. (77) vanish identically. The second-order terms are evaluated by inserting between each pair of operators a set of basis states complete except for the vectors $|\alpha; p\rangle$ and $|\beta; p + 1\rangle$ (which are eliminated by the complementary projection operator Q). The elements then reduce to

$$\mathcal{R}_{11}^{(1;2)} = -\mathcal{R}_{22}^{(1;2)} = \tfrac{1}{2}(\delta\Omega + i\delta\Gamma), \qquad (78a)$$

in which

$$\delta\Omega = \mathrm{Re}\{\mathcal{R}_{11}^{(1;2)} - \mathcal{R}_{22}^{(1;2)}\} = \frac{2V^2\Omega'}{\Gamma^2 + \Omega'^2} \qquad (78b)$$

and

$$\delta G = \mathrm{Im}\{\mathcal{R}_{11}^{(1;2)} - \mathcal{R}_{22}^{(1;2)}\} = \frac{2V^2\Gamma}{\Gamma^2 + \Omega'^2} \qquad (78c)$$

are precisely the shifts derived in chapter 2 from the OFT approach, Eqs. (2.35) and (2.36). [Recall that $\Gamma = \tfrac{1}{2}(\gamma_\alpha - \gamma_\beta)$ and $\Omega' = \omega + \omega_0$.]

Let us now consider r-photon transitions between states $|\alpha; p\rangle$ and $|\beta; p + r\rangle$ (with $r > 1$). Setting $q = p + r$ and evaluating the elements of the resolvent matrix $\mathcal{G}(E)$, Eq. (76), we find that the lowest nonvanishing term in the infinite series expansion [Eq. (67)] for the nondiagonal elements of $\mathcal{R}$ is the rth-order term

$$R_{12}^{(r;r)} = R_{21}^{(r;r)} = \frac{\sqrt{\dfrac{(p+r)!}{p!\bar{n}^r}}V^r}{(E - E_{\beta; p+1})(E - E_{\alpha; p+2})\cdots(E - E_{\alpha; p+r-1})}. \qquad (79a)$$

At the anticrossing point, $E_{\alpha;\,p} = E_{\beta;\,p+r} \equiv E_{ac}$, the off-diagonal element $\mathcal{R}_{12}^{(r;\,r)}$ is a slowly varying nonsingular function of the integration variable E if, as is assumed to be the case, m.q. processes are well separated. Making the approximation that $\mathcal{R}_{12}^{(r;\,r)}(E) \approx \mathcal{R}_{12}^{(r;\,r)}(E_{ac})$ and assuming that the mean number of photons in the field $\bar{n} \approx p$, we can reduce Eq. (79a) to the compact expression

$$\mathcal{R}_{12}^{(r;\,r)} = \frac{V^r}{(r-1)!!\;\omega^{(r-1)/2}\;\displaystyle\prod_{j=1,\,3,\,5,\,...}^{r-2}\left(\Omega_{(j)} + i\Gamma\right)} \tag{79b}$$

in which

$$\Omega_{(j)} = j\omega - \omega_0. \tag{80}$$

The double factorial function $x!!$ is defined by the relation

$$x!! = x \cdot (x-2) \cdot (x-4) \cdots \begin{matrix} 1 & (x \text{ odd}) \\ 2 & (x \text{ even}) \end{matrix}. \tag{81}$$

To evaluate the diagonal elements of $\mathcal{R}$ it is sufficient to retain the first nonzero term in the series (65) which is second order in $\mathcal{H}$. These elements take the form

$$\mathcal{R}_{11}^{(r;\,2)} = -\mathcal{R}_{22}^{(r;\,2)} = \tfrac{1}{2}\left(-\delta\Omega_{(2)} + i\delta\Gamma_{(2)}\right) \tag{82a}$$

with

$$\delta\Omega_{(2)} = 2V^2\left(\frac{\Omega_{(1)}}{\Gamma^2 + \Omega_{(1)}^2} + \frac{\Omega_{(-1)}}{\Gamma^2 + \Omega_{(-1)}^2}\right) \tag{82b}$$

and

$$\delta\Gamma_{(2)} = 2V^2\Gamma\left(\frac{1}{\Gamma^2 + \Omega_{(1)}^2} + \frac{1}{\Gamma^2 + \Omega_{(-1)}^2}\right). \tag{82c}$$

Note that the difference in sign between the frequency shift of Eq. (82a) and that of Eq. (78a) (for a single-quantum transition) is due merely to a generalization of our previous notation. From Eq. (80) it follows that $\Omega_{(1)} = \Omega$ and $\Omega_{(-1)} = -\Omega'$; the $\delta\Omega$ of Eq. (78b) is then seen to correspond to the second term of $\delta\Omega_{(2)}$ in Eq. (82b). The new feature in Eqs. (82b, c) is the term containing $\Omega_{(1)}$ that arises from inclusion of the state $|\beta;\,p+1\rangle$ in the set of basis states comprising the complementary projection operator Q; this state had to be excluded from Q in the treatment of single-quantum transitions.

Substitution of the elements of $\mathcal{R}$ into the contour integral (68) leads to the amplitudes

$$c_{\alpha;\,p}^{(r)}(t) = e^{-i(\omega_\alpha + p\omega - \frac{1}{2}i\gamma_\alpha)t}\, e^{(\Gamma - i\Omega_{(r)})t/2}$$

$$\times \left[\cosh \nu_{(r)}t - \frac{(\Gamma - \delta\Gamma_{(2)}) - i(\Omega_{(r)} - \delta\Omega_{(2)})}{\nu_{(r)}} \sinh \nu_{(r)}t \right],$$

$$\tag{83a}$$

$$c_{\beta;\,p+r}^{(r)}(t) = e^{-i(\omega_\beta + (p+r)\omega - \frac{1}{2}i\gamma_\beta)t}\, e^{-(\Gamma - i\Omega_{(r)})t/2}$$

$$\times \left(\frac{-i\mathcal{R}_{21}^{(r;r)}}{\nu_{(r)}} \right) \sinh \nu_{(r)}t \tag{83b}$$

with

$$\nu_{(r)} = \sqrt{\tfrac{1}{4}\left[(\Gamma - \delta\Gamma_{(2)}) - i(\Omega_{(r)} + \delta\Omega_{(2)})\right]^2 - (\mathcal{R}_{12}^{(r;r)})^2}. \tag{83c}$$

This generalized notation shows clearly the similarity in form of the m.q. probability amplitudes to the equations derived from the RWA theory in chapters 2 and 3. The energy separation of the two participating levels at their anticrossing is $\mathrm{Im}\{\nu_{(r)}\}_{(\Omega_{(r)}=0)}$. If the appropriate expressions are used for $\delta\Omega_{(2)}$ and $\delta\Gamma_{(2)}$, then Eqs. (79a) through (83c) apply to single- as well as to multiple-quantum transitions—that is, for all $r \geq 1$. Specialized to the case of a triple-quantum process, Eqs. (79b) and (83c) become

$$\mathcal{R}_{12}^{(3;3)} = \frac{V^3}{2\omega(\Omega_{(1)} + i\Gamma)}, \tag{84a}$$

$$\nu_{(3)} = \sqrt{\tfrac{1}{4}\left[(\Gamma - \delta\Gamma_{(2)}) - i(\Omega_{(2)} + \delta\Omega_{(2)})\right]^2 - \frac{V^6}{4\omega^2(\Omega_{(1)} + i\Gamma)^2}}. \tag{84b}$$

4.8 ONE- AND THREE-PHOTON LINESHAPES

Although it is conceptually useful to identify for analysis multiple-quantum transitions of a particular order, it is often the case experimentally that more than one such order of transition can occur at the applied rf frequency. It is instructive, therefore, to examine a triple-quantum lineshape not only in isolation (as resolvent theory of an anticrossing region enables us to do), but also as part of an overall lineshape (as deduced by the six-state analysis of section 4.6) to which the one-photon transition is the dominant contribution. As an example consider transitions between the states $|a\rangle = |4^2S_{1/2}00\rangle$ and $|\beta\rangle = |4^2P_{1/2}10\rangle$ of hydrogen. The energy separation (Lamb shift) for purposes of calculation is here taken to be $\omega_0/2\pi = 120$ MHz (more accurately, it is 114.5 MHz); state decay rates are $\gamma_\alpha = 4.35 \times 10^6$ s^{-1} and $\gamma_\beta = 80.7 \times 10^6$ s^{-1}.

Figures 4.7 and 4.8 respectively show the variation in one- and three-photon lineshapes for several interaction strengths V and fixed interaction time of 50 ns. The single-photon transition probability, Eq. (57a), was computed by numerical solution of the Schrödinger equation in the six-state approximation discussed in section 4.6. It is to be noted that, while resonance profiles broaden substantially with increased field stength, the shift of the line center is slight. Secondary oscillations build up in intensity, and the lineshape near resonance alternates between a maximum and minimum.

For the same range of field strengths, the behavior of the triple-photon profile is quite different. The spectra increase rapidly in intensity with increased coupling strength accompanied by significant shifts in the line center and only slight broadening. At $V = 10$ MHz the single-photon lineshape has already split and shows marked secondary oscillations. Oscillatory behavior of comparable degree in the triple-photon lineshape does not set in until about $V = 50$ MHz, as illustrated in figure 4.9.

The three-photon profiles of figure 4.8 were obtained in two ways: (a) by numerical solution of the Schrödinger equation in the six-state approximation [Eq. (57b)], and (b) from the analytical expressions [Eqs. (83a, b)] for two coupled states in the region of a triple-quantum anti-crossing. For low rf field strengths the agreement between the two methods is good, but as the field strength increases, the two-state calculation overestimates the resonance probability and level shift. The reason for this is clear from the energy level diagram (figure 4.4) and discussion of section 4.7. Application of a stronger field increases the "repulsion" between coupled levels with the consequence that several anticrossing regions (corresponding to different multiphoton processes) begin to merge. Such occurrences are not accounted for in the two-state calculation. Nevertheless, for the highest coupling strength ($V = 30$ MHz) shown in figure 4.8, the triple-quantum transition probability at exact resonance

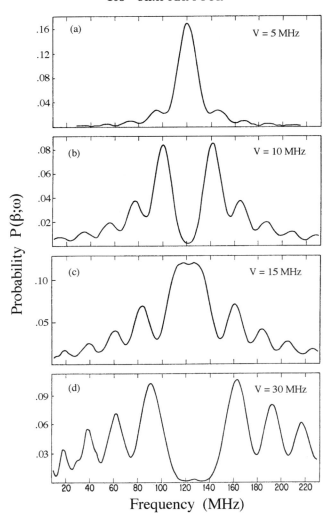

Figure 4.7 Variation of single-quantum lineshapes with interaction strength for an interaction time of 50 ns. The atomic states $|\alpha\rangle = |4^2S_{1/2}00\rangle$ and $|\beta\rangle = |4^2P_{1/2}10\rangle\rangle$ decay at rates $\gamma_\alpha = 4.35 \times 10^6$ s^{-1}, $\gamma_\beta = 80.7 \times 10^6$ s^{-1}; the level separation is taken to be $\omega_0/2\pi = 120$ MHz.

obtained from the two-state approximation deviates from that of the six-state approximation only by a factor of two.

Unlike multiple-quantum transitions observed in traditionally field-swept magnetic resonance experiments, for which the resonance locations vary as integral multiples of the Larmor frequency, the locations

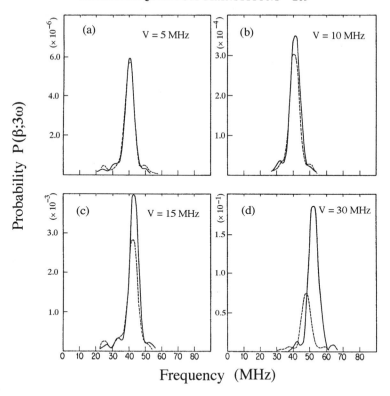

Figure 4.8 Variation of triple-quantum lineshapes with interaction strength for same parameters as in figure 4.7. Dashed lines indicate six-state approximation; solid lines indicate two-state approximation in region of level anticrossing.

of multiphoton electric resonance transitions vary reciprocally with the order of the process. This is perhaps one reason why they have been diffi-cult to detect. For a given rf field strength, the transition probability falls rapidly with increasing level separation. For small level separations, how-ever, the multiphoton lineshapes overlap each other and the dominant single-quantum profile. From Eqs. (83a–c) one also sees that, except for an overall exponential decay factor, only the decay rate difference enters into the transition probability. Thus, magnetic resonance spectra result-ing from the coupling of two states of the same lifetime are as narrow as if the levels were completely stable. Electric resonance lineshapes, however, are the result of transitions between unstable states of differ-

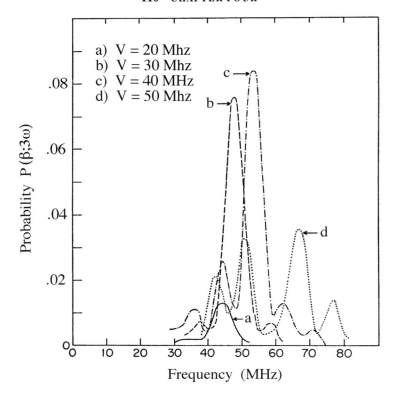

Figure 4.9 Comparison of triple-quantum lineshapes for large rf field strengths V. Parameters are the same as for figure 4.7.

ent lifetimes and are correspondingly broader, a feature that exacerbates the problem of lineshape overlap.

Three-photon rf electric dipole transitions can be observed in coupled states of lifetimes sufficiently different to allow preparation of an initially pure beam of the longer-lived state and optical detection of radiative decay from the shorter-lived state. The level separation should not be so large as to result in a negligible three-photon transition probability, but large enough so that the lineshape is not completely obscured by the single-photon resonance profile. Also, there should be no overlap from other fine structure transitions in the same manifold. As an example, figure 4.10 compares the $4^2S_{1/2}(00)$–$4^2P_{1/2}(10)$ total resonance profile calculated from Eq. (47) with the corresponding lineshape for a single-photon transition alone. For an interaction time of 50 ns and coupling

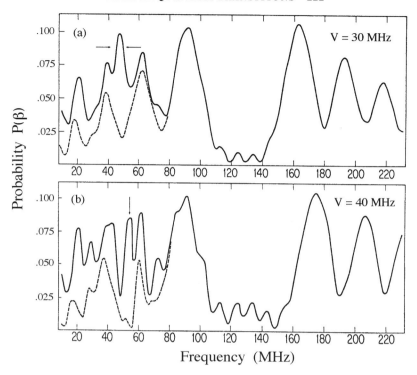

Figure 4.10 Total lineshape (solid line) including single- and triple-quantum spectra for $V = 30$ and 40 MHz. Three-photon resonances are marked by arrows. The dashed portions of the spectra show the single-photon background in the vicinity of the triple-quantum transition. Parameters are the same as for figure 4.7.

strengths $V = 30$ MHz and 40 MHz (experimental parameters readily produced in the laboratory), the three-photon transitions indicated by arrows are clearly visible against the single-photon background.

APPENDIX 4A: SEMICLASSICAL THEORY OF MULTIPHOTON TRANSITIONS

The discreteness intrinsic to the quantum field formalism can be incorporated into the semiclassical formalism by means of Fourier analysis, a perspective which we examine in this appendix.

Starting with Eq. (2.7) and performing the phase transformation of Eq. (35),

$$c_\mu(t) = a_\mu(t)e^{-i(\omega_\mu + \lambda_\mu)t} \quad (\mu = \alpha, \beta), \qquad (A1)$$

with $\lambda_\alpha = -\omega_0$ and $\lambda_\beta = 0$ [which follow from the standard energy scale; see Eqs. (52a, b) with $p = k = 0$], we obtain the two semiclassical equations

$$\frac{da_\alpha(t)}{dt} = -\left(i\omega_0 + \tfrac{1}{2}\gamma_\alpha\right)a_\alpha(t) - iV\left(e^{i\omega t} + e^{-i\omega t}\right)\alpha_\beta(t), \qquad (A2a)$$

$$\frac{da_\beta(t)}{dt} = -\tfrac{1}{2}\gamma_\beta a_\beta(t) - iV\left(e^{i\omega t} + e^{-i\omega t}\right)a_\alpha(t). \qquad (A2b)$$

Upon expansion of the two amplitudes in Fourier series,

$$a_\alpha(t) = \sum_{n=-\infty}^{\infty} a_\alpha^{(2n)} e^{i2n\omega t}, \qquad (A3a)$$

$$a_\beta(t) = \sum_{n=-\infty}^{\infty} a_\beta^{(2n+1)} e^{i(2n+1)\omega t}, \qquad (A3b)$$

substitution of the series into Eqs. (A2a, b), and collection of terms containing exponential phase factors $e^{i\omega t}$ of the same power, there results the infinite set of equations

$$\frac{da_\alpha^{(2n)}}{dt} = -i\left(\omega_0 + 2n\omega - \tfrac{1}{2}i\gamma_\alpha\right)a_\alpha^{(2n)} - iV\left(a_\beta^{(2n+1)} + a_\beta^{(2n-1)}\right), \qquad (A4a)$$

$$\frac{da_\beta^{(2n+1)}}{dt} = -i\left((2n+1)\omega - \tfrac{1}{2}i\gamma_\beta\right)a_\beta^{(2n+1)} - iV\left(a_\alpha^{(2n)} + a_\alpha^{(2n+2)}\right), \qquad (A4b)$$

which are seen to follow from Eqs. (37) and (41) derived by quantizing the rf field.

In the foregoing analysis the amplitudes are labeled with a field "quantum number" as well as the state of the atom. Although in theory only the atom is quantized, the solution for the Fourier components does permit a proper evaluation of individual multiphoton transition probabilities under the circumstance that the field contains a large number of photons (that is, that the field is classical) so that the radicals appearing in the exact set of quantum field equations (24) may all be set to unity.

Semiclassical theory (but not the fully quantized field theory) fails, however, in the description of processes, such as spontaneous emission, for which the field contains only one or a few photons.

APPENDIX 4B: RESOLVENTS, PROPAGATORS,
AND GREEN'S FUNCTIONS

The formal solution to the Schrödinger equation $\mathcal{H}\Psi(t) = i\partial\Psi(t)/\partial t$ can be expressed in terms of a unitary time-evolution operator

$$\Psi(t) = U(t, t_0)\Psi(t_0), \tag{B1}$$

which satisfies the differential equation

$$\mathcal{H}U(t, t_0) - i\frac{\partial U(t, t_0)}{\partial t} = 0 \tag{B2}$$

with initial condition

$$U(t_0, t_0) = \mathbf{1} \tag{B3}$$

(where $\mathbf{1}$ is the unit operator). When the hamiltonian is independent of time, Eq. (B2) is readily integrated to yield $U(t, t_0) = e^{-i\mathcal{H}(t-t_0)}$ [which also satisfies Eq. (B3)]. In general, Eq. (B2) is not directly integrable, and alternative ways to deduce $U(t, t_0)$ are needed.

Equation (59) asserts that the time-evolution operator is equivalent to the contour integral

$$U(t, t_0) = \frac{1}{2\pi i} \oint_C dE\, e^{-iE(t-t_0)} G(E). \tag{B4}$$

To demonstrate that this is indeed the case, it is sufficient to show that the representation (B4) obeys the differential equation (B2) and initial condition (B3) expected of the time-evolution operator. From the definition of the resolvent, Eq. (60), it follows that $(E - \mathcal{H})G(E) = \mathbf{1}$; applied to the integrand of (B4), this operator identity leads to

$$\mathcal{H}U(t, t_0) = \frac{1}{2\pi i} \oint_C dE\, e^{-iE(t-t_0)} EG(E) - \frac{1}{2\pi i} \oint_C e^{-iE(t-t_0)}\, dE$$
$$= i\frac{\partial U(t, t_0)}{\partial t}. \tag{B5}$$

In the above expression, the second integral vanishes identically (the integrand contains no poles), and the first integral is readily seen to be the indicated time derivative. Thus, Eq. (B4) satisfies the equation of motion (B2).

If the initial condition is to be satisfied, then the integral relation

$$\frac{1}{2\pi i} \oint_C dE\, G(E) = \mathbf{1} \tag{B6}$$

must hold at $t = t_0$. To prove this, project Eq. (B6) between the eigenstates $\langle \mu |$ and $| \nu \rangle$ of $\mathcal{H}$ to obtain

$$\frac{1}{2\pi i} \oint_C dE\, G(E)_{\mu\nu} = \frac{1}{2\pi i} \oint_C \frac{\delta_{\mu\nu}\, dE}{E_\mu - E} = \delta_{\mu\nu} \tag{B7a}$$

or equivalently

$$\frac{1}{2\pi i} \oint_C \frac{dE}{E_\mu - E} = 1, \tag{B7b}$$

the validity of which follows immediately from the Cauchy residue theorem.

Although mention of the resolvent appears much less widely in standard quantum mechanics textbooks than that of the propagator, the two operators are closely related. In brief, the (retarded) propagator satisfies an equation of motion

$$\mathcal{H} G^{(+)}(t - t') - i \frac{\partial G^{(+)}(t - t')}{\partial t} = -\delta(t - t') \tag{B8}$$

[with $\delta(t - t')$ the Dirac delta function] and "propagates" the state vector forward in time according to the relation

$$\Psi(t) = \int_{-\infty}^{+\infty} G^{(+)}(t - t')\Psi(t')\, dt', \tag{B9}$$

which is suggestive of Huygens's principle in optics. Note, however, that Eq. (B9) is an operator relation; projected onto coordinate basis states, the state vectors become wavefunctions, and the coordinate matrix elements of the propagator become Green's functions.

The resolvent operator may be thought of as a Fourier transform of the propagator,

$$G^{(+)}(t) = \frac{1}{2\pi} \int_{-\infty}^{+\infty} dE \, e^{-iEt} G^{(+)}(E), \tag{B10}$$

from which follows the form in which it is often found:

$$G^{(+)}(E) = \lim_{\eta \to 0}(E - \mathcal{H} + i\eta)^{-1}. \tag{B11}$$

Equation (B11) is operationally equivalent to Eq. (60) which defines the resolvent $G(E)$. In Eq. (B11), however, E and the eigenvalues of $\mathcal{H}$ are real valued; the imaginary infinitesimal $i\eta$ ($\eta > 0$) serves to depress the eigenvalues below the real axis so that a causally consistent contour can be constructed. Mathematically, the real E-axis is a cut in the Riemann plane, and $G^{(+)}(E)$ is the limit of the resolvent when the cut is approached from above. Similarly, the Fourier transform $G^{(-)}(E)$ of the advanced Green's function is the limit of the resolvent when the cut is approached from below. In applying the resolvent, Eq. (60), to the problem of transitions between atomic states of finite lifetime, we did not need to concern ourselves with the foregoing limiting procedure. Each eigenvalue of $\mathcal{H}$ was intrinsically complex valued with a negative imaginary part deriving from the phenomenological decay rates. Thus, the poles of $G(E)$ automatically fell in the lower half of the complex plane.

NOTES

1. The angular momentum of a photon with helicity $+1$ is parallel to the wave vector (or linear momentum). In the convention of optical physics this state corresponds to left circular polarization (LCP) since the electric vector of the light rotates toward the left side of an observer facing the light source. Conversely, the angular and linear momenta of a photon with helicity -1 are antiparallel; this state corresponds to right circular polarization (RCP). Optical and microwave engineers often define the handedness of light with respect to an observer facing the direction of the light beam, that is, away from the source, in which case the connection between helicity and circular polarization is reversed.

2. P. Kusch, *Phys. Rev.* **93** (1954):1022; **101** (1955):627.

3. J. Brossel, B. Cagnac, and A. Kastler, *J. Phys. et Rad.* **15** (1954):6.

4. J. Margerie and J. Brossel, *Comptes Rendus* **241** (1955):373.

5. M. P. Silverman, *Optical Electric Resonance Investigation of a Fast Atomic Beam* (Ph.D. Thesis, Harvard University, 1973), chap. 5.

6. See, for example, E. Merzbacher, *Quantum Mechanics*, 2nd edition (Wiley, New York, 1970), chap. 20.

7. Although retaining two multiplicative minus signs may seem cumbersome, the form of Eq. (7) shows explicitly the sign intrinsic to the interaction and the sign arising from the charge of the electron.

8. R. Glauber, *Phys. Rev.* **131** (1963):2766.

9. The photon number state $|n\rangle$ is actually a nonclassical state in that it corresponds to no solution to Maxwell's electromagnetic equations. The distinctions between the coherent state $|a\rangle$ and the photon number state are manifested in their degrees of second-order (and higher) coherence, functions that do not play a role in the phenomena examined in this book. Such properties are observed in correlation experiments like those pioneered by Hanbury Brown and Twiss (HBT). See R. Loudon, *The Quantum Theory of Light*, 2nd edition (Oxford University Press, New York, 1983), chap. 6 and M. P. Silverman, *More than One Mystery: Explorations of Quantum Interference* (Springer, New York, 1995), chap. 3 for discussions of HBT-type experiments with light and electrons, respectively.

10. M. Göppert-Mayer, *Ann. Physik* **9** (1931):273.

11. P. I. Richards, *Phys. Rev.* **73** (1947):254.

12. E. A. Power and S. Zienau, *Phil. Trans. Roy. Soc. (London)* **A251** (1959):427.

13. The equivalence of expressions (34b) and (34c), which employ direct product states $|\mu; n\rangle$, to Eqs. (2) and (3) is readily established by factoring out summations over atomic states or field states and applying the completeness relation $\sum_{\mu} |\mu\rangle\langle\mu| = \sum_{n} |n\rangle\langle n| = \mathbf{1}$, where $\mathbf{1}$ is the unit operator.

14. H. Wieder and T. G. Eck, *Phys. Rev.* **153** (1967):103.

15. M. L. Goldberger and K. M. Watson, *Collision Theory* (Wiley, New York, 1964), chap. 8.

16. A good discussion of residues and the evaluation of integrals by complex analysis can be found in H. K. Crowder and W. W. McCuskey, *Topics in Higher Analysis* (Macmillan, New York, 1964), chap. 5.

17. L. Mower, *Phys. Rev.* **142** (1966):799.

CHAPTER 5

The Decay of Coupled States

5.1 PERSPECTIVES ON RADIATION DAMPING

In general, the usual starting point for the analysis of electric (and magnetic) resonance spectroscopy is the nonrelativistic Schrödinger equation from which are derived occupation and transition probabilities as functions of time, frequency, and field strength. In the preceding chapters we examined the theory of rf-coupled states from various vantage points, treating the applied field both as a classical perturbation of a quantized atomic system and as a quantized part of a fully quantum mechanical atom-field system. There remains, however, the question of accounting for radiative decay.

The detection of electric and magnetic resonance transitions between unstable atomic states is often accomplished indirectly by monitoring the change in spontaneous emission from these states as their populations are altered by an applied rf field. In atomic physics this procedure underlies a variety of precision experiments designed to measure fine and hyperfine level separations, the interpretation of which is contingent upon a consistent and accurate theory of the observed spectral lineshapes. In experimental approaches that employ a resonance cell ("bottle" methods), excitation, spectroscopy, and detection all occur within the same region, and it is important to ascertain what effects the oscillating field has on the temporal and spectral properties of radiative decay. As pointed out in the preface, one advantage of the use of a fast atomic beam is that the spectroscopy and detection regions are spatially separated, and therefore the observed radiation is emitted from field-free states. But spontaneous emission, even if not detected, can occur all throughout the state-selection and spectroscopy regions where atoms are subjected to rf fields, and it is therefore necessary to know here too whether coupled transient states evolve in accordance with the decay law assumed to hold in the absence of rf fields.

The process of spontaneous emission is ordinarily introduced into the Schrödinger equation by a phenomenological Hamiltonian ($\mathcal{H}_D$) whose (imaginary) eigenvalues represent the state decay rates and which, when acting alone, gives rise to an exponential decrease in time of the state amplitudes. This is the approach that we have followed to this point. The theoretical validity of the exponential decay law for spontaneous emission by an atom in a field-free environment was established around 1930 by Weisskopf and Wigner[1] and has subsequently been demonstrated by others using mathematical methods of increased rigor and sophistication.[2] As is well known from both classical and quantum mechanics, the frequency distribution resulting from exponential damping is Lorentzian. Nevertheless, the validity of these results may be questioned for the case where spontaneous transitions take place in the presence of an external time-dependent field. This is the problem we address in this chapter.

For a basic understanding of the pertinent physical processes, which involve optical transitions induced by fluctuations of the electromagnetic vacuum, it is necessary to quantize the optical radiation field; quantization of the rf field, however, is now an unnecessary complication. In this chapter, therefore, we again consider excited atomic states coupled by a classical monochromatic rf field and, after applying the phase transformation of the previous chapter to remove all time dependence from the equations of motion except that intrinsic to the state amplitudes, derive a set of equations that can be solved by the resolvent operator method.

Although to anticipate matters removes any suspense, it is perhaps best to state at the outset that the previous chapters were not in vain; the phenomenological approach to damping does indeed result as a good approximation from a quantum treatment of the vacuum electromagnetic field. Beyond this demonstration, the analysis of the present chapter enables us to study the effects of the rf field on the resulting optical lineshapes.

5.2 THE QUANTIZED OPTICAL FIELD

In figure 5.1 is shown a level diagram of the model sytem to be described. Atomic states $|1\rangle$ and $|2\rangle$ (with energies $\hbar\omega_1 > \hbar\omega_2$ and Bohr

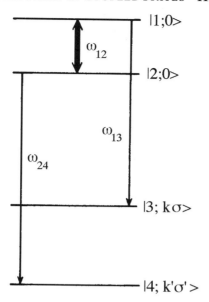

Figure 5.1 Radiative decay from two excited states (Bohr frequency ω_{12}) coupled by an applied rf field of frequency ω.

frequency $\omega_{12} = \omega_1 - \omega_2$), coupled by an oscillating field of frequency ω, can decay via the emission of a single photon to lower states $|3\rangle$ and $|4\rangle$, respectively (with $\omega_3 > \omega_4$). As discussed in the previous chapter, the quantized electromagnetic field may be described by photon number states $|n_{k\sigma}\rangle$, where $n_{k\sigma}$ is the number of photons having momentum $\hbar k$, energy $\hbar ck \equiv \hbar\omega_k$, and helicity $\sigma = +1$ (for LCP) and -1 (for RCP). In contrast to the previous chapter, however, where only a single mode of the quantized rf field was considered in the treatment of multiple-quantum transitions (and energy and polarization labels were consequently superfluous), many modes of the quantized optical field contribute to radiative decay.

For a noninteracting atom and radiation field, the evolution in time of the total system is governed by a Hamiltonian of the form of Eq. (4.1),

$$\mathcal{H}_0 = \mathcal{H}_a + \mathcal{H}_r, \tag{1a}$$

except that now the atomic Hamiltonian has only real eigenvalues ω_μ corresponding to state energies (in angular frequency units). The field Hamiltonian (where r signifies optical radiation, not rf) acts upon the

n-photon state $|n_{\mathbf{k}\sigma}\rangle$ to generate the eigenvalue $n\omega_k$. Thus, the energy of the total system, representable by the direct product of atom and field basis states, is given by

$$\mathcal{H}_0|\mu; n_{\mathbf{k}\sigma}\rangle = (\omega_\mu + n\omega_k)|\mu; n_{\mathbf{k}\sigma}\rangle. \tag{1b}$$

The interaction $\mathcal{H}_1$ coupling the atom to the radiation field, expressed earlier by Eq. (4.7), may be written more usefully in the expanded form

$$\mathcal{H}_1 = -\frac{(-e)}{c}\sum_{\mathbf{k}\sigma}(a_{\mathbf{k}\sigma}\mathbf{j}_{-\mathbf{k}} \cdot \hat{\mathbf{e}}_\sigma(\mathbf{k}) + a_{\mathbf{k}\sigma}^\dagger\mathbf{j}_{\mathbf{k}} \cdot \hat{\mathbf{e}}_\sigma^*(\mathbf{k})), \tag{2}$$

where $\mathbf{j}_{\mathbf{k}}$ is the kth Fourier component

$$\mathbf{j}_{\mathbf{k}} = \int d^3r\, e^{-i\mathbf{k}\cdot\mathbf{r}}\mathbf{J}(\mathbf{r}) \tag{3a}$$

of the operator

$$\mathbf{J}(\mathbf{r}) = \frac{1}{2m}\sum_i(\mathbf{p}_i\delta(\mathbf{r} - \mathbf{r}_i) + \delta(\mathbf{r} - \mathbf{r}_i)\mathbf{p}_i) \tag{3b}$$

corresponding to the electron current density (number of electrons per volume) in the atom. The operator $\delta(\mathbf{r} - \mathbf{r}_i)$ in Eq. (3b) is the three-dimensional Dirac delta function (which, technically, is not a function but what mathematicians term a distribution) with dimension of inverse volume. Here $\mathbf{p}_i$ and $\mathbf{r}_i$ are respectively the linear momentum and coordinate operators of the ith electron; $\mathbf{r}$ is not an operator, but the point in space (field point) at which the current is evaluated. Since momentum and coordinate operators do not commute, the correct expresssion for $\mathbf{J}(\mathbf{r})$ must be symmetric in the ordering of $\mathbf{p}_i$ and $\delta(\mathbf{r} - \mathbf{r}_i)$.

The operator coefficients $a_{\mathbf{k}\sigma}^\dagger$ and $a_{\mathbf{k}\sigma}$ in a Fourier series expansion of the vector potential field

$$\mathbf{A}(\mathbf{r}) = \sum_{\mathbf{k}\sigma}(a_{\mathbf{k}\sigma}\hat{\mathbf{e}}_\sigma(\mathbf{k})e^{i\mathbf{k}\cdot\mathbf{r}} + a_{\mathbf{k}\sigma}^\dagger\hat{\mathbf{e}}_\sigma^*(\mathbf{k})e^{-i\mathbf{k}\cdot\mathbf{r}}) \tag{4}$$

act on photon basis states to create and annihilate, respectively, a photon of momentum $\hbar\mathbf{k}$, helicity σ, and polarization vector $\hat{\mathbf{e}}_\sigma(\mathbf{k})$. The matrix elements for these processes, a generalization of Eqs. (4.5a,b), are

$$\langle n_{\mathbf{k}\sigma} - 1|a_{\mathbf{k}\sigma}|n_{\mathbf{k}\sigma}\rangle = \left(\frac{2\pi c^2}{\omega_k \mathcal{V}}\right)^{1/2} \sqrt{n_{\mathbf{k}\sigma}}, \tag{5a}$$

$$\langle n_{\mathbf{k}\sigma} + 1|a_{\mathbf{k}\sigma}^\dagger|n_{\mathbf{k}\sigma}\rangle = \left(\frac{2\pi c^2}{\omega_k \mathcal{V}}\right)^{1/2} \sqrt{n_{\mathbf{k}\sigma} + 1}. \tag{5b}$$

Note that the coefficient α of the preceding chapter [Eq. (4.8b)] is now no longer a constant amplitude in the expression for $\mathbf{A}(\mathbf{r})$, but a k-dependent factor incorporated in the matrix elements (5a,b) instead. As before, $\mathcal{V}$ is the volume of the enclosure within which the electromagnetic field is quantized. The interaction Hamiltonian (2) again omits a term proportional to $\mathbf{A} \cdot \mathbf{A}$ since this term does not contribute to processes involving creation or annihilation of single-photon states. In this regard, we will use the simpler notation $|\mathbf{k}\sigma\rangle$, rather than $|1_{\mathbf{k}\sigma}\rangle$, to represent single-photon states.

The coupling of the atom with the rf field, a classical single-mode electromagnetic field of frequency ω and amplitude $\mathbf{E}_0$, is effected by the electric dipole interaction

$$\mathcal{H}_2 = -\boldsymbol{\mu}_E \cdot \mathbf{E}_0 \cos \omega t \tag{6}$$

as first introduced in Eq. (2.1). Here, as in chapter 2, we will make use of the rotating-field approximation and neglect the counter-rotating component since this approximation leads to exactly soluble equations whose solutions are satisfactory enough for our present purposes. At some arbitrary time t after the initiation of the interaction at $t = 0$ the total atom-field system is represented by a superposition of basis states

$$|\Psi(t)\rangle = c_1(t)|1; 0\rangle + c_2(t)|2; 0\rangle + \sum_{\mathbf{k}\sigma} c_{3;\mathbf{k}\sigma}(t)|3; \mathbf{k}\sigma\rangle$$
$$+ \sum_{\mathbf{k}'\sigma'} c_{4;\mathbf{k}'\sigma'}(t)|4; \mathbf{k}'\sigma'\rangle \tag{7}$$

with time-dependent coefficients to be determined from the Schrödinger equation (2.4). The initial conditions are chosen so that at time $t = 0$ only state $|1; 0\rangle$ is populated. Since the photon momentum is actually a continuously distributed variable, the sum over radiation field modes in Eq. (7) is implemented through the relation

$$\sum_{\mathbf{k}\sigma} \rightarrow \sum_{\sigma} \int \eta(\omega_k)\, d\Omega_k\, d\omega_k, \tag{8a}$$

in which

$$\eta(\omega_k) = \frac{\mathscr{V}\omega_k^2}{(2\pi c)^3} \tag{8b}$$

is the field mode density per unit solid angle Ω_k.[3]

5.3 STATE AMPLITUDES AND RADIATIVE DECAY RATES

To solve the Schrödinger equation, we proceed in the now familiar way by making a phase transformation [Eq. (4.35)]

$$|\Psi(t)\rangle = e^{-i(\mathscr{H}_0 + \Lambda)t}|\Psi_\Lambda(t)\rangle \tag{9}$$

in which Λ is a diagonal operator in the space of states $\{|\mu; \mathbf{k}\sigma\rangle\}$ with eigenvalues $\{\lambda_{\mu;\mathbf{k}\sigma}\}$. Since $\mathscr{H}_0$ and Λ commute, Eq. (9) is equivalent to two sequential transformations (in either order). As written, the first transformation (to the interaction representation) removes high-frequency terms (that is, terms containing optical frequencies), whereas the second transformation resets the reference energy level to remove time-dependent phase factors and thereby render a set of coupled differential equations with time-independent coefficients. Leaving the details of these calculations, which are somewhat tedious, to the original literature,[4] we can summarize the results to this point as follows. With the eigenvalues of Λ defined by

$$\lambda_{1;0} = -\lambda_{2;0} = \tfrac{1}{2}(\omega - \omega_{12}) \equiv \tfrac{1}{2}\Omega_{12}, \tag{10a}$$

$$\lambda_{3;\mathbf{k}\sigma} = \tfrac{1}{2}\Omega_{12} - \Omega_{13;k} \quad (\text{with } \Omega_{13;k} \equiv \omega_k - \omega_{13}), \tag{10b}$$

$$\lambda_{4;\mathbf{k}'\sigma'} = -\tfrac{1}{2}\Omega_{12} - \Omega_{24;k'} \quad (\text{with } \Omega_{24;k'} \equiv \omega_{k'} - \omega_{24}) \tag{10c}$$

(we employ throughout the standard notation $\omega_{ij} \equiv \omega_i - \omega_j$), the amplitudes $b_{\mu;\mathbf{k}\sigma} \equiv \langle \mu; \mathbf{k}\sigma \mid \Psi_\Lambda \rangle$ are found to satisfy the following set of equations:

$$\frac{db_{1;0}}{dt} = \tfrac{1}{2}i\Omega_{12}b_{1;0} - iV_{12}b_{2;0} - i\sum_{k\sigma}\mathcal{H}_{13;k\sigma}b_{3;k\sigma}, \tag{11a}$$

$$\frac{db_{2;0}}{dt} = -iV_{12}b_{1;0} - \tfrac{1}{2}i\Omega_{12}b_{2;0} - i\sum_{k'\sigma'}\mathcal{H}_{24;k'\sigma'}b_{4;k'\sigma'}, \tag{11b}$$

$$\frac{db_{3;k\sigma}}{dt} = -i\mathcal{H}_{13;k\sigma}^{\dagger}b_{1;0} + i\left(\tfrac{1}{2}\Omega_{12} - \Omega_{13;k}\right)b_{3;k\sigma}, \tag{11c}$$

$$\frac{db_{4;k'\sigma'}}{dt} = -i\mathcal{H}_{24;k'\sigma'}^{\dagger}b_{2;0} - i\left(\tfrac{1}{2}\Omega_{12} + \Omega_{24;k'}\right)b_{4;k'\sigma'}. \tag{11d}$$

The matrix element of the rf interaction is

$$(\hbar)V_{12} = \langle 1| - \tfrac{1}{2}\boldsymbol{\mu}_E \cdot \mathbf{E}_0|2\rangle, \tag{12a}$$

and the matrix elements coupling states through the optical field are

$$(\hbar)\mathcal{H}_{13;k\sigma} = -\frac{(-e)}{c}\sqrt{\frac{2\pi(\hbar)c^2}{\omega_k\mathcal{V}}}\langle 1|\mathbf{j}_{-\mathbf{k}}\cdot\hat{\mathbf{e}}_{\sigma}(\mathbf{k})|3\rangle, \tag{12b}$$

$$(\hbar)\mathcal{H}_{24;k'\sigma'} = -\frac{(-e)}{c}\sqrt{\frac{2\pi(\hbar)c^2}{\omega_{k'}\mathcal{V}}}\langle 2|\mathbf{j}_{-\mathbf{k'}}\cdot\hat{\mathbf{e}}_{\sigma'}(\mathbf{k'})|4\rangle. \tag{12c}$$

Appropriate factors of $\hbar$ have been inserted in parentheses to restore units of energy.

Although the preceding set of equations has a physically meaningful solution only for $t > 0$, it will prove convenient to extend the time domain to include the negative time axis as well. In so doing, however, one must account for the singularity of $b_{1;0}$ at $t = 0$ (it jumps from 0 at $t = 0_-$ to 1 at $t = 0_+$) by adding to Eq. (11a) a Dirac delta function $\delta(t)$ (since the first derivative of a step function is a delta function). This enables us to make Fourier transformations

$$b_{\mu;k\sigma}(t) = -\frac{1}{2\pi i}\int_{-\infty}^{+\infty}dE\, e^{-iEt}G_{\mu;k\sigma}(E), \tag{13a}$$

$$i\delta(t) = -\frac{1}{2\pi i}\int_{-\infty}^{+\infty}dE\, e^{-iEt} \tag{13b}$$

to convert the infinite set of differential equations (11a–d) into a set of algebraic equations for the kernels $G_{\mu;\mathbf{k}\sigma}(E)$:

$$(E + \tfrac{1}{2}\Omega_{12})G_{1;0}(E) = V_{12}G_{2;0}(E) + \sum_{\mathbf{k}\sigma} \mathscr{H}_{13;\mathbf{k}\sigma} G_{3;\mathbf{k}\sigma}(E) + 1, \quad (14a)$$

$$(E - \tfrac{1}{2}\Omega_{12})G_{2;0}(E) = V_{12}G_{1;0}(E) + \sum_{\mathbf{k}'\sigma'} \mathscr{H}_{24;\mathbf{k}'\sigma'} G_{4;\mathbf{k}'\sigma'}(E), \quad (14b)$$

$$(E - \Omega_{13;k} + \tfrac{1}{2}\Omega_{12})G_{3;\mathbf{k}\sigma}(E) = \mathscr{H}^{\dagger}_{13;\mathbf{k}\sigma} G_{1;0}(E), \quad (14c)$$

$$(E - \Omega_{24;k'} - \tfrac{1}{2}\Omega_{12})G_{4;\mathbf{k}'\sigma'}(E) = \mathscr{H}^{\dagger}_{24;\mathbf{k}'\sigma'} G_{2;0}(E). \quad (14d)$$

Although executed a little differently than before, one recognizes in this method (due to Heitler[5]) the essential features of the resolvent operator formalism (as elucidated in appendix B of chapter 4).

Solution of equations (14a–d) with implementation of the initial conditions leads to the rather complicated expressions

$$G_{1;0}(E) = \frac{E - \tfrac{1}{2}\Omega_{12} + \tfrac{1}{2}i\gamma_2(E)}{\left(E + \tfrac{1}{2}\Omega_{12} + \tfrac{1}{2}i\gamma_1(E)\right)\left(E - \tfrac{1}{2}\Omega_{12} + \tfrac{1}{2}i\gamma_2(E)\right) - V_{12}^2},$$
$$(15a)$$

$$G_{2;0}(E) = \frac{V_{12}}{\left(E + \tfrac{1}{2}\Omega_{12} + \tfrac{1}{2}i\gamma_1(E)\right)\left(E - \tfrac{1}{2}\Omega_{12} + \tfrac{1}{2}i\gamma_2(E)\right) - V_{12}^2},$$
$$(15b)$$

$$G_{3;\mathbf{k}\sigma}(E) = \frac{\mathscr{H}^{\dagger}_{13;\mathbf{k}\sigma}\left(E - \tfrac{1}{2}\Omega_{12} + \tfrac{1}{2}i\gamma_2(E)\right)\zeta\left(E - \Omega_{13;k} + \tfrac{1}{2}\Omega_{12}\right)}{\left(E + \tfrac{1}{2}\Omega_{12} + \tfrac{1}{2}i\gamma_1(E)\right)\left(E - \tfrac{1}{2}\Omega_{12} + \tfrac{1}{2}i\gamma_2(E)\right) - V_{12}^2},$$
$$(15c)$$

$$G_{4;\mathbf{k}'\sigma'}(E) = \frac{\mathscr{H}^{\dagger}_{24;\mathbf{k}'\sigma'}V_{12}\zeta\left(E - \Omega_{24;k'} - \tfrac{1}{2}\Omega_{12}\right)}{\left(E + \tfrac{1}{2}\Omega_{12} + \tfrac{1}{2}i\gamma_1(E)\right)\left(E - \tfrac{1}{2}\Omega_{12} + \tfrac{1}{2}i\gamma_2(E)\right) - V_{12}^2},$$
$$(15d)$$

in which the functions $\gamma_1(E)$ and $\gamma_2(E)$ are defined by

$$\tfrac{1}{2}i\gamma_1(E) = \sum_{\mathbf{k}\sigma} \left|\mathscr{H}^{\dagger}_{13;\mathbf{k}\sigma}\right|^2 \zeta\left(E - \Omega_{13;k} + \tfrac{1}{2}\Omega_{12}\right), \quad (16a)$$

$$\tfrac{1}{2}i\gamma_2(E) = \sum_{\mathbf{k}'\sigma'} \left|\mathscr{H}^{\dagger}_{24;\mathbf{k}'\sigma'}\right|^2 \zeta\left(E - \Omega_{24;k'} + \tfrac{1}{2}\Omega_{12}\right). \quad (16b)$$

Note that Eqs. (14c) and (14d) do not automatically lead to unique solutions for $G_{3;\mathbf{k}\sigma}$ and $G_{4;\mathbf{k}'\sigma'}$ since to each kernel can be added a delta function multiplied by an arbitrary coefficient.[6] Setting this coefficient to $-i\pi$ uniquely determines the wave amplitudes as the set obeying the stipulated initial conditions. This procedure accounts for the appearance in Eqs. (15c, d) of Dirac zeta functions defined by

$$\zeta(x) = \lim_{t \to \infty} \frac{1 - e^{ixt}}{x} = \mathcal{P}(1/x) - i\pi\delta(x) \tag{17}$$

where $\mathcal{P}$ signifies the Cauchy principal value.

It is instructive to examine at this point the significance of the γ-functions of Eqs. (16a, b). Consider $\gamma_1(E)$; from Eqs. (16a) and (17) we can write

$$\gamma_1(E) = 2\pi \sum_{\mathbf{k}\sigma} |\mathcal{H}^\dagger_{13;\mathbf{k}\sigma}|^2 \delta\left(E - \Omega_{13;k} + \tfrac{1}{2}\Omega_{12}\right)$$

$$+ 2i \sum_{\mathbf{k}\sigma} \mathcal{P} \frac{|\mathcal{H}^\dagger_{13;\mathbf{k}\sigma}|^2}{E - \Omega_{13;k} + \tfrac{1}{2}\Omega_{12}}. \tag{18}$$

The first term, $\mathrm{Re}\{\gamma_1(E)\}$, according to first-order time-dependent perturbation theory (the "Fermi golden rule"), is the total probability of transition per unit time out of atomic state $|1\rangle$, and leads to a spectral line of finite width. Correspondingly, the second term, $\mathrm{Im}\{\gamma_1(E)\}$, represents a level shift due to the emission and reabsorption of virtual photons.

To evaluate the integrals (13a) with the kernels of Eqs. (15a–d) by contour integration, it is necessary that $\gamma_1(E)$ and $\gamma_2(E)$ be analytic in the whole complex plane with the exception of a branch cut along the real axis. These functions do have this property although it will not be demonstrated here. The poles in the lower half of the second Riemann sheet give rise to exponential decay, whereas the branch line gives rise to corrections to exponential decay. Since the deviations from exponential decay occur only for time intervals very short or very long compared with the mean lifetime—indeed clear evidence for the nonexponential decay of a quantum system has been observed for the first time only recently[7]— we can obtain the major contribution to the integrals by approximating $\gamma_1(E)$ and $\gamma_2(E)$ by their values at $E = 0$, or equivalently at the respective optical frequencies $\omega_k = \omega_{13} + \tfrac{1}{2}\Omega_{12}$ and $\omega_{k'} = \omega_{24} - \tfrac{1}{2}\Omega_{12}$.

In that case the integrand has no branch line, and the integral can be simply evaluated by deforming the contour into a semicircular path in the lower half plane as implemented in chapter 4 (see figure 4.6). The constants $\mathrm{Re}\{\gamma_1(0)\} \equiv \gamma_1$ and $\mathrm{Re}\{\gamma_2(0)\} \equiv \gamma_2$ are identifiable as the familiar phenomenological decay rates (or inverse lifetimes) of states $|1\rangle$ and $|2\rangle$. Since we are not concerned in this chapter with calculating QED level shifts, we will assume in what follows that the corresponding imaginary parts $\mathrm{Im}\{\gamma_1(0)\} = \Delta\omega_1$ and $\mathrm{Im}\{\gamma_2(0)\} = \Delta\omega_2$ have been absorbed in the eigenfrequencies ω_1 and ω_2 respectively.[8]

Upon substitution of kernels (15a) and (15b) into the integral (13a) and application of the theory of residues (the two poles of the integrand lie in the lower half of the complex plane), we obtain amplitudes $b_{1;0}(t)$ and $b_{2;0}(t)$ in complete agreement with those of the RWA theory [Eq. (2.23) with $c_1^0 = 1$ and $c_2^0 = 0$] derived by application of a phenomenological decay Hamiltonian $\mathscr{H}_D$ and direct temporal integration of the time-dependent Schrödinger equation. From these amplitudes follow the occupation and transition probabilities that we have examined before, but which we now write in the alternative form

$$P_1(\omega, t) = e^{-\bar{\gamma}t}\left|\cos\mu t - (\Gamma - i\Omega_{12})\frac{\sin\mu t}{2\mu}\right|^2, \qquad (19a)$$

$$P_2(\omega, t) = e^{-\bar{\gamma}t}\left|V_{12}\frac{\sin\mu t}{\mu}\right|^2 \qquad (19b)$$

with $\bar{\gamma} = \frac{1}{2}(\gamma_1 + \gamma_2)$ the mean decay rate, $\Gamma = \frac{1}{2}(\gamma_1 - \gamma_2)$, and $\mu = iv = \sqrt{\frac{1}{4}(\Omega_{12} + i\Gamma)^2 + V_{12}^2}$ [cf. Eq. (2.24f)] the factor corresponding to the precession frequency of the classical vector model.

Of particular interest in the present chapter are the amplitudes, obtained by integration of kernels (15c) and (15d),

$$b_{3;\mathbf{k}\sigma} = e^{-i(\Omega_{13;k} - \frac{1}{2}\Omega_{12})t}\frac{H_{13;\mathbf{k}\sigma}^{\dagger}}{2\mu}$$
$$\times \left[\frac{[\Gamma - i(\Omega_{12} - 2\mu)]\left[e^{-\frac{1}{2}\bar{\gamma}t}e^{i(\Omega_{13;k} - \mu - \frac{1}{2}\Omega_{12})t} - 1\right]}{\bar{\gamma} - 2i(\Omega_{13;k} - \mu - \frac{1}{2}\Omega_{12})}\right.$$
$$\left. - \frac{[\Gamma - i(\Omega_{12} + 2\mu)]\left[e^{-\frac{1}{2}\bar{\gamma}t}e^{i(\Omega_{13;k} + \mu - \frac{1}{2}\Omega_{12})t} - 1\right]}{\bar{\gamma} - 2i(\Omega_{13;k} + \mu - \frac{1}{2}\Omega_{12})}\right],$$
$$(20a)$$

$$b_{4;\,\mathbf{k}'\sigma'} = e^{-i(\Omega_{24;\,k'} - \frac{1}{2}\Omega_{12})t} \frac{H^{\dagger}_{24;\,\mathbf{k}'\sigma'}}{2\mu}$$

$$\times \left[\frac{e^{-\frac{1}{2}\bar{\gamma}t} e^{i(\Omega_{24;\,k'} - \mu + \frac{1}{2}\Omega_{12})t} - 1}{\bar{\gamma} - 2i(\Omega_{24;\,k'} - \mu + \frac{1}{2}\Omega_{12})} - \frac{e^{-\frac{1}{2}\bar{\gamma}t} e^{i(\Omega_{24;\,k'} + \mu + \frac{1}{2}\Omega_{12})t} - 1}{\bar{\gamma} - 2i(\Omega_{24;\,k'} + \mu + \frac{1}{2}\Omega_{12})} \right]$$

$$(20b)$$

that lead to final states with the emission of optical radiation. Each amplitude is linearly proportional to a matrix element for spontaneous emission and can therefore be written as the product of the optical matrix element (which contains the angular degrees of freedom and polarization of the field mode) and a function dependent on time and the interaction with the rf field (but independent of photon polarization and wave-vector orientation)

$$b_{3;\,\mathbf{k}\sigma}(t) = H^{\dagger}_{13;\,\mathbf{k}\sigma} \beta_{3;\,k}(t), \qquad (21a)$$

$$b_{4;\,\mathbf{k}'\sigma'}(t) = H^{\dagger}_{24;\,\mathbf{k}'\sigma'} \beta_{4;\,k'}(t). \qquad (21b)$$

Explicit expressions for the β-functions can be read directly from Eqs. (20a, b). The photon frequency distribution functions are then derived by summing [by means of Eq. (8a)] the transition probabilities $P_{3;\,\mathbf{k}\sigma} = |b_{3;\,\mathbf{k}\sigma}|^2$ and $P_{4;\,\mathbf{k}'\sigma'} = |b_{4;\,\mathbf{k}'\sigma'}|^2$ over all radiation field modes at the frequencies ω_k and $\omega_{k'}$, respectively, to obtain

$$P_3(\omega_k, t) = \frac{\gamma_1}{2\pi} |\beta_{3;\,k}|^2, \qquad (22a)$$

$$P_4(\omega_{k'}, t) = \frac{\gamma_2}{2\pi} |\beta_{4;\,k'}|^2. \qquad (22b)$$

5.4 EMISSION LINESHAPES

The course of spontaneous decay in time and as a function of applied rf field strength and frequency is contained in the distribution functions (22a, b). As an example, let us consider the properties of the emission spectra arising from transitions induced between two excited states of

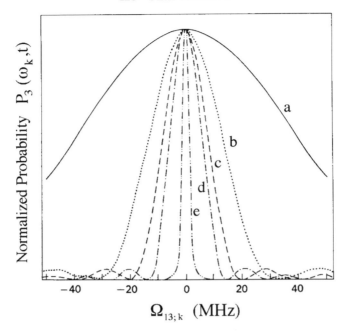

Figure 5.2 Normalized emission lineshapes for radiative decay from initially populated state $|1\rangle$ for resonant transitions ($\Omega_{12} = 0$) by a weak rf field ($V_{12} = 2$ MHz) at increasing times t (in ns): (a) 10; (b) 30; (c) 50; (d) 70; (e) ∞. The excited state decay rates are $\gamma_1 = 6.25 \times 10^6$ s^{-1} and $\gamma_2 = 185.2 \times 10^2$ s^{-1}.

very different lifetimes, such as the hydrogen $3S$ and $3P$ states, which decay predominantly to $2P$ and $1S$, respectively. The mean lifetimes of the rf-coupled states are $\tau_{3S} = 160$ ns, $\tau_{3P} = 5.4$ ns, and the corresponding decay rates $\gamma_{3S} \equiv \gamma_1 = 6.25 \times 10^6$ s^{-1}, $\gamma_{3P} \equiv \gamma_2 = 185.2 \times 10^6$ s^{-1}. It is assumed that the longer-lived $3S$ state (higher in energy than $3P$ by approximately $\omega_{12}/2\pi = 300$ MHz) is initially populated and the 3P state is initially empty.

Figure 5.2 shows the emission profile of the $3S$ state—that is, the probability $P_3(\omega_k, t)$ as a function of $\Omega_{13;k}$ (the difference in photon frequency from the Bohr frequency ω_{13})—at increasing times during the interaction with a resonant rf field ($\Omega_{12} = 0$) of relatively weak strength ($V_{12} = 2$ MHz). The lineshapes are normalized to the same value at their maximum point ($\Omega_{13;k} = 0$) to facilitate comparison of spectral widths. For interaction times of extremely short duration, $\bar{\gamma}t \ll 1$, $V_{12}t \ll 1$,

Eq. (22a) reduces to

$$\lim_{t \to 0} P_3(\omega_k, t) = \frac{8\gamma_1 t^2}{\pi}, \tag{23}$$

which is independent of $\Omega_{13;k}$, and thus yields a profile of infinite width. With increasing interaction time, the profiles narrow, approaching, as t approaches infinity, a spike of width γ_{min}, where γ_{min} is the smaller of γ_1 and γ_2. In the limit of an infinitely long interaction time, by which interval the excited states have assuredly decayed, Eq. (22a) reduces to

$$P_3(\omega_k, \infty) = \frac{\gamma_1}{8\pi|\mu|^2} \left| \frac{[\Gamma - i(\Omega_{12} - 2\mu)]}{\bar{\gamma} - 2i(\Omega_{13;k} - \mu - \frac{1}{2}\Omega_{12})} \right.$$
$$\left. - \frac{[\Gamma - i(\Omega_{12} + 2\mu)]}{\bar{\gamma} - 2i(\Omega_{13;k} + \mu - \frac{1}{2}\Omega_{12})} \right|^2. \tag{24a}$$

The corresponding distribution function for the $3P$ state to which Eq. (22b) reduces is

$$P_4(\omega_{k'}, \infty) = \frac{\gamma_2 V_{12}^2}{8\pi|\mu|^2} \left| \frac{1}{\bar{\gamma} - 2i(\Omega_{24;k'} - \mu + \frac{1}{2}\Omega_{12})} \right.$$
$$\left. - \frac{1}{\bar{\gamma} - 2i(\Omega_{24;k'} + \mu + \frac{1}{2}\Omega_{12})} \right|^2. \tag{24b}$$

As one check of the theory for conditions where the outcome is well known, note that in the absence of the rf field ($V_{12} = 0$), $P_4(\omega_{k'}, t)$ vanishes (since state $|2\rangle$ remains unpopulated and therefore cannot radiate) and the resulting radiation distribution from decay of state $|1\rangle$,

$$P_3(\omega_k, \infty) = \frac{\gamma_1}{2\pi} \frac{1}{(\Omega_{13;k})^2 + \frac{1}{4}\gamma_1^2}, \tag{25}$$

reproduces, as expected, the Lorentzian lineshape derived by Weisskopf and Wigner for the case of field-free spontaneous emission.

A graphic portrayal of the buildup of emission probability with time is illustrated in figure 5.3 which shows the relative intensity of the emission profiles, again under the condition of a weak resonant rf field. For a time interval (beginning with the excitation to the $3S$ state) short with respect to the lifetime τ_{3S}, the wings of the profile develop first. As

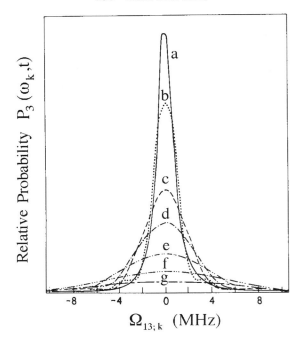

Figure 5.3 Relative emission probability P_3 as a function of time (in ns) for resonant ($\Omega_{12} = 0$) transitions induced by a weak rf field ($V_{12} = 2$ MHz): (a) ∞; (b) 400; (c) 200; (d) 150; (e) 100; (f) 70; (g) 50. Same decay rates as in figure 5.2.

time increases, the proportion of radiation from the center of the line increases and constitutes the major part of the emitted light when $t > \tau_{3S}$. An analogous set of lineshapes is exhibited in figure 5.4 for the case of a much larger coupling strength, $V_{12} = 15$ MHz. Although for short interaction times the profiles build up in much the same manner as those of figure 5.3, the later time development is substantially different, the lineshapes becoming double peaked. Let us look more closely at the origin of these features.

Figure 5.5 shows the variation in emission profile $P_3(\omega_k, \infty)$ at resonance ($\Omega_{12} = 0$) with increasing values of the rf interaction V_{12}. At the lowest field strength shown (curve a), the condition $\Gamma^2 \gg 4V_{12}^2$, and the line is a narrow Lorentzian centered about $\Omega_{13;k} = 0$ with width of the order of $\gamma_{\min}$. As the coupling between the excited states increases, the radiation profile broadens and finally splits into two narrow lines symmetrically displaced about $\Omega_{13;k} = 0$. To understand this, we return

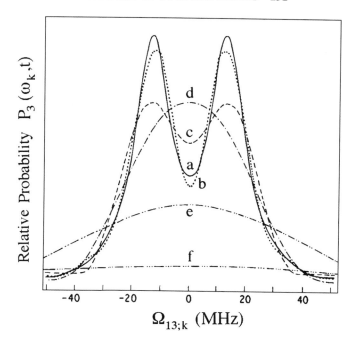

Figure 5.4 Strong-field ($V_{12} = 15$ MHz) relative emission profiles P_3 for resonant transitions ($\Omega_{12} = 0$) at interaction times t (in ns): (a) ∞; (b) 60 ns; (c) 40 ns; (d) 30 ns; (e) 10 ns; (f) 5 ns. Same decay rates as in figure 5.2.

to Eq. (24a), which can be reexpressed as

$$P_3(\omega_k, \infty) = \frac{2\gamma_1\gamma_2^2}{16\bar{\gamma}^2\pi}\left(\frac{1 + (2\Omega_{13;k}/\gamma_2)^2}{(\Omega_{13;k})^2 + \frac{1}{4}\left[\dfrac{\gamma_1\gamma_2 + V_{12}^2 - 4\Omega_{13;k}^2}{\bar{\gamma}}\right]^2}\right). \quad (26)$$

For field strengths that satisfy $\Gamma^2 \gg 4V_{12}^2$, the frequency dependence is determined primarily by the first term in the denominator, whereupon the expression reduces approximately to the form of a Lorentzian line-shape,

$$P_3(\omega_k, \infty) \approx \text{constant} \times \left(\frac{1}{(\Omega_{13;k})^2 + \frac{1}{4}\gamma_v^2}\right) \quad \text{(weak coupling)}, \quad (27)$$

with field-dependent width $\gamma_v = (\gamma_1\gamma_2 + V_{12}^2)/(\gamma_1 + \gamma_2)$. If $\gamma_1 \gg \gamma_2$ or $\gamma_2 \gg \gamma_1$, then γ_v reduces to $\gamma_{min} + V_{12}^2/\gamma_{max}$ in agreement with curve a of

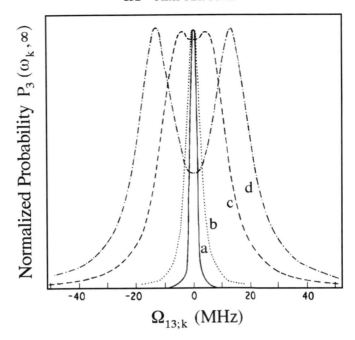

Figure 5.5 Variation in emission lineshape P_3 with coupling strength for resonant transitions ($\Omega_{12} = 0$) between the radiating states at an infinitely long time after excitation. $V_{12} =$ (a) 2 MHz; (b) 5 MHz; (c) 10 MHz; (d) 15 MHz. Same decay rates as in figure 5.2.

figures 5.2 and 5.5. As V_{12} increases, the width increases, but eventually a point is reached where one can no longer ignore the other terms in $(\Omega_{13;k})^2$ appearing in Eq. (26).

When $4V_{12}^2 \gg \Gamma^2$, we see from Eq. (24a) that each denominator separately passes through a minimum determined by $\Omega_{13;k} = \mp V_{12}\sqrt{1 - \Gamma^2/4V_{12}^2}$ in which the upper sign refers to the first term in Eq. (24a) and the lower sign refers to the second term. When one term reaches its maximum, the other is substantially damped, thus permitting one to approximate $P_3(\omega_k, \infty)$ by a sum of two Lorentzian functions,

$$
\begin{aligned}
P_3(\omega_k, \infty) \\
\approx \frac{\gamma_1}{8\pi}\left(1 + \frac{\Gamma^2}{4\mu^2}\right)&\left(\frac{1}{(\Omega_{13;k} + \mu)^2 + \frac{1}{4}\bar{\gamma}^2} + \frac{1}{(\Omega_{13;k} - \mu)^2 + \frac{1}{4}\bar{\gamma}^2}\right) \\
&\text{(strong coupling),} \qquad (28)
\end{aligned}
$$

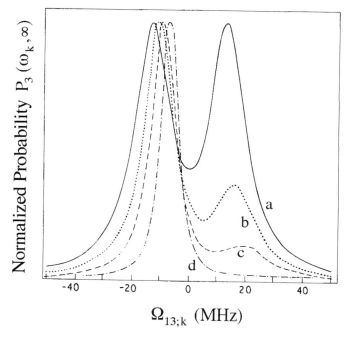

Figure 5.6 Variation in strong-field ($V_{12} = 15$ MHz) emission lineshape as a function of detuning frequency $\Omega_{12} =$ (a) 0 MHz; (b) 5 MHz; (c) 10 MHz; (d) 20 MHz. Same decay rates as in figure 5.2.

centered about $\mp\mu$, respectively, with $\mu = V_{12}\sqrt{1 - \Gamma^2/4V_{12}^2}$, in agreement with curve d of figure 5.5.

Away from resonance ($\Omega_{12} \neq 0$), the emission spectrum becomes distorted and one of the two components, depending on the sign of Ω_{12}, diminishes rapidly with increasing $|\Omega_{12}|$, while the peak of the other component is shifted by $\frac{1}{2}\Omega_{12}$. This behavior is shown in figure 5.6.

With regard to spectroscopic methods that employ optical detection, the effect on the emission probability of the rf frequency ω is of particular importance. Figures 5.7 and 5.8 show the changes in $P_3(\omega_k, \infty)$ and $P_4(\omega_{k'}, \infty)$ at the center of each respective optical transition as one sweeps the frequency ω (or, equivalently, the detuning frequency Ω_{12}) through resonance for several values of the rf field strength. Although the profiles broaden as the rf coupling V_{12} gets larger, they remain ostensibly Lorenzian in shape, in marked contrast to the behavior exhibited by the spectral profiles of figure 5.5. To understand this behavior

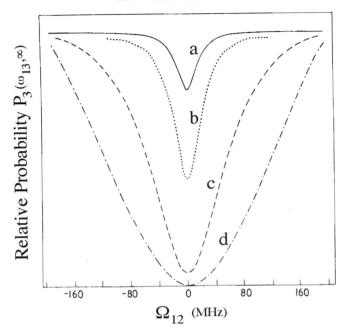

Figure 5.7 Variation with coupling strength of the optically detected rf resonance signal ($\Omega_{13;k} = 0$) from the (initially populated) upper excited state. $V_{12} =$ (a) 1 MHz; (b) 2 MHz; (c) 5 MHz; (d) 10 MHz. Same decay rates as in figure 5.2.

and obtain a quantitative estimate of the spectral widths we substitute $\Omega_{13;k} = \Omega_{24;k'} = 0$ into Eqs. (24a,b), which then reduce to

$$P_3(\omega_{13}, \infty) = \frac{\gamma_2^2}{2\pi\gamma_1}\left(\frac{1 + (2\Omega_{12}/\gamma_2)^2}{\Omega_{12}^2 + \frac{1}{4}\left(\gamma_2 + \frac{4V_{12}^2}{\gamma_1}\right)^2}\right), \qquad (29a)$$

$$P_4(\omega_{24}, \infty) = \frac{4V_{12}^2}{2\pi\gamma_2}\left(\frac{1}{\Omega_{12}^2 + \frac{1}{4}\left(\gamma_1 + \frac{4V_{12}^2}{\gamma_2}\right)^2}\right). \qquad (29b)$$

Equations (29a) and (29b) yield spectral profiles symmetric about $\Omega_{12} = 0$ with widths $\gamma_2 + 4V_{12}^2/\gamma_1$ and $\gamma_1 + 4V_{12}^2/\gamma_2$, respectively. For γ_2 larger

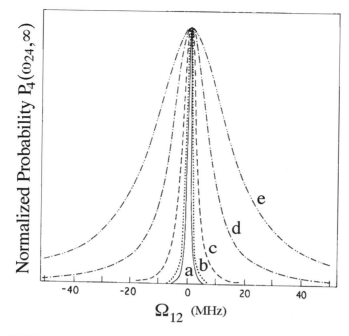

Figure 5.8 Variation with coupling strength of the optically detected rf resonance signal ($\Omega_{24;k'} = 0$) from the (initially unpopulated) lower excited state. $V_{12} =$ (a) 1 MHz; (b) 2 MHz; (c) 5 MHz; (d) 10 MHz; (e) 15 MHz. Same decay rates as in figure 5.2.

than the range of Ω_{12} required to sweep across the line, Eq. (29a) is a good approximation to a Lorentzian profile.

One last feature of the decay process worth considering, because of the seemingly paradoxical issue it raises, is the evolution in time of the spontaneous emission from the lower (or upper) excited state. Figure 5.9 shows this variation in $P_4(\omega_{24}, t)$ at resonance ($\Omega_{12} = 0$) for several values of the coupling strength. We note first that, as V_{12} increases, the intensity peaks earlier, the emission probability increasing gradually and monotonically as long as the condition $\Gamma^2 > 4V_{12}^2$ is satisfied. However, under the reverse condition, $4V_{12}^2 > \Gamma^2$, the rise is steep and exhibits a weak transient modulation before assuming a steady-state value. The mathematical origin of these results becomes apparent when the resonant conditions $\Omega_{12} = \Omega_{24;k'} = 0$ are substituted into Eq. (22b) to yield

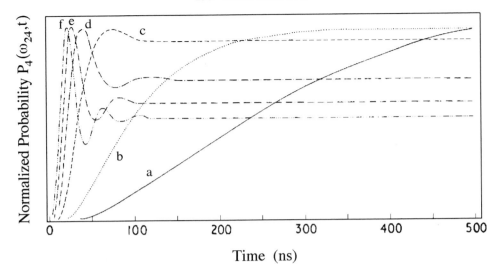

Figure 5.9 Time evolution of the resonant ($\Omega_{12} = 0$) spontaneous emission at frequency ω_{24} for different values of the rf coupling strength $V_{12} =$ (a) 2 MHz; (b) 5 MHz; (c) 10 MHz; (d) 15 MHz; (e) 20 MHz; (f) 25 MHz. Same decay rates as in figure 5.2.

the simpler expression

$$P_4(\omega_{24}, t) = \frac{\gamma_2 V_{12}^2}{2\pi\mu^2(4V_{12}^2 + \gamma_1\gamma_2)^2}$$

$$\times \left(e^{-\frac{1}{2}\bar{\gamma}t}(\bar{\gamma}\sin\mu t + 2\mu\cos\mu t) - \mu\right)^2, \quad (30)$$

in which $\mu = \sqrt{V_{12}^2 - \frac{1}{4}\Gamma^2}$. For coupling strength V_{12}^2 greater than $\frac{1}{4}\Gamma^2$, μ is real valued and the probability $P_4(\omega_{24}, t)$ oscillates in time.

At first thought it might seem strange that the emission probability is modulated, for this would appear to indicate an rf-induced depopulation of the *ground state* and back coupling to the excited states. Since $P_4(\omega_{k'}, t)$ is the probability of finding the atom in the ground state $|4\rangle$ at time t if it was in the excited state $|1\rangle$ at $t = 0$, $P_4(\omega_{k'}, t)$ is proportional to the number of photons $|k'\sigma'\rangle$ emitted over the interval 0 to t.

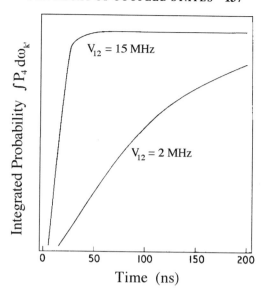

Figure 5.10 Time variation of the total frequency-integrated emission probability $\int_0^\infty P_4(\omega_{k'}, t)\, d\omega_{k'}$ at resonance $(\Omega_{12} = 0)$. The spectra are normalized to their maximum values at $t = \infty$; the ratio of maximum values of the 2 MHz to 15 MHz plots is about 1:3.6. Same decay rates as in figure 5.2.

Any decrease in this number would suggest that radiative decay in the presence of an external rf field is not completely reversible.

That there is no paradox here is demonstrated in figure 5.10 which shows, as a function of time, the total area under each emission spectrum. These frequency-integrated profiles increase steadily and eventually level off when the major proportion of excited states has decayed; they exhibit no ground state depopulation regardless of rf coupling strength or frequency. Thus, the total radiative decay into all modes of the vacuum electromagnetic field, whether in a region of field-free space or in the presence of a rf perturbation coupling the excited states, is still an irreversible process. The oscillations shown in figure 5.9—and which also occur in an analogous plot (not shown) of $P_3(\omega_k, t)$—signify a temporal modulation of the number of ground state atoms formed in a narrow photon frequency interval, but no modulation of the total ground state population. This function, as expected, continuously increases in time, ultimately reaching a steady-state value.[9]

NOTES

1. V. Weisskopf and E. Wigner, *Z. Phys.* **63** (1930):54; **65** (1930):18.

2. W. Heitler and S. T. Ma, *Proc. Irish Acad. Sci.* **52** (1949):109; E. Arnous and S. Zienau, *Helv. Phys. Acta.* **24** (1951):279; F. Low, *Phys. Rev.* **88** (1952):53.

3. The number of electromagnetic field modes in a volume $d^3\mathbf{r}d^3\mathbf{p}$ of phase space is $2d^3\mathbf{r}d^3\mathbf{p}/h^3 = 2d^3\mathbf{r}(p^2\,dp\,d\Omega)/h^3$, where the factor 2 comes from the two allowable polarizations for each frequency ω and wave vector $\mathbf{k}$. Dividing this number by the spatial volume $d^3\mathbf{r}$ and differential solid angle $d\Omega$ and setting $p = \hbar k = h\omega/2\pi c$ leads to the mode density η of Eq. (8b).

4. M. P. Silverman and F. M. Pipkin, *J. Phys. B: Atom. Molec. Phys.* **5** (1972):2236.

5. W. Heitler, *The Quantum Theory of Radiation* (Oxford University Press, London, 1954), chap. V.

6. As a consequence of the identity $x\delta(x) = 0$, the solution $g(x)$ to an equation of the form $xg(x) = h(x)$, where $h(x)$ is nonsingular at $x = 0$, can always be written as $g(x) = h(x)[(1/x) + c\delta(x)]$ with c an arbitrary constant.

7. P. F. Schewe, *Physics Today* **50** No. 8 (August 1997):8.

8. The calculation of such self-energy terms was a problematic issue in the early development of quantum electrodynamics, but is now a routine part of advanced quantum theory; see, for example, J. J. Sakurai, *Advanced Quantum Mechanics* (Addison-Wesley, Reading, MA, 1967), pp. 70–72, 292–294.

9. An analogous phenomenon is that of "quantum beats" in which the spontaneous emission from two (or more) states impulsively excited (for example, by pulsed laser) is modulated in time at the frequency corresponding to their energy difference. The effect is manifested only in polarized components of the decay radiation, and vanishes when all polarizations are detected. See M. P. Silverman, *More than One Mystery: Explorations of Quantum Interference* (Springer, New York, 1995), chap. 4 for discussion of laser-induced quantum beats.

CHAPTER 6

Optical Detection Theory

6.1 THE PROCESS OF DETECTION

As the previous chapters have shown, the location, width, and shape of resonance curves obtained by detection of spontaneous emission from excited atomic states coupled by a frequency-swept rf field can reveal much about atomic level structure and excitation. In the accelerator-based method of probing atoms that is the focus of attention of this book, a fast atomic beam, generated by charge-exchange collisions between an accelerated ion beam and a stationary gas or foil target, passes rapidly (before the states of interest have decayed) out of the rf-spectroscopy region into a separate optical detection chamber.

The spontaneous emission of radiation from decaying particles, which may or may not have been coherently excited, and the subsequent detection of this radiation subject to constraints of detector efficiency and geometry, constitute a widely encountered problem throughout experimental physics. In nuclear physics, particularly with regard to the angular distribution of correlated photons, the subject of γ-ray detection has received an extensive theoretical treatment in terms of the statistical or density operator and the apparatus efficiency matrix.[1] These two concepts provide an elegant and effective formalism with which to analyze atomic physics experiments as well.[2]

The effect of a photon-monitoring apparatus on the combined atom–radiation field system can be represented by a detection operator $P^{(d)}$ which projects out of the totality of available states only those to which the detector responds. The sensitivity of response is given by the elements of an efficiency matrix generated by the expansion of $P^{(d)}$ in a suitable complete set of basis states. By evaluating the trace of the product of $P^{(d)}$ with the statistical operator $\rho^{(a;r)}$ of the entire atom-radiation system, one can calculate the intensity, polarization, and angular distribution of the decay radiation as a function of the initial elements $\rho_{ii'}$ of the atomic density matrix $\rho^{(a)}$ and the variables of the radiofrequency field.

The elements of the statistical operator $\rho^{(a)}$ characterizing a beam of atoms that traverses an oscillatory field and decays in a region of field-free space were derived in chapter 3. By applying or generalizing relations of the preceding chapter, one can also construct the appropriate density matrix elements for "bottle" experiments in which both the rf coupling and spontaneous decay occur in the same region. In the absence of rf fields, the theory developed in the present chapter can be used to analyze atomic lifetime measurements.

We begin by considering the atom in the optical detection chamber (and no longer under the influence of the rf field). Prior to decay, the atom and radiation field constitute two independent subsystems whose complete initial description (to the extent permitted by quantum mechanics) is contained in the statistical operator $\rho^{(a;r)} = \rho^{(a)} \otimes \rho^{(r)}$, which is the direct product of the initial density operators of the atom and radiation subsystems alone; the symbol $\otimes$ here refers to the outer or Kronecker product.

A representation of $\rho^{(a)}$ is obtained by expanding the statistical operator in a complete set of the eigenvectors $\{|i\rangle\}$ of the field-free atomic Hamiltonian $\mathcal{H}_a$ with eigenvalues $\{\omega_i\}$:

$$\rho^{(a)} = \sum_{i,\,i'} |i\rangle \rho_{ii'} \langle i'|. \tag{1}$$

Recall from the previous chapter that the eigenvalues of $\mathcal{H}_a$ are real-valued energies (or frequencies) and do not include the phenomenological decay rates. In the representation (1), the label i of a state vector connotes the entire set of coupled angular momentum (c.a.m.) quantum numbers needed to specify the state uniquely; for the hydrogen atom these quantum numbers are n, L, J, F, and M_F. The intrinsic electron and nuclear spin quantum numbers ($S = I = \frac{1}{2}$ for hydrogen) are the same for all atomic states and need not be written explicitly unless required for clarity.

In the absence of external fields, the atomic density matrix elements evolve in time according to Eq. (3.7), reexpressed now as

$$\rho_{ii'}(t) = e^{-(i\omega_{ii'} + \frac{1}{2}\gamma_{ii'})(t - t_0)} \rho_{ii'}(t_0), \tag{2}$$

where $\omega_{ii'}$ and $\gamma_{ii'}$ are defined by Eqs. (3.8) and (3.9). Note that, in contrast to Eq. (3.7), the initial matrix elements $\rho_{ii'}(t_0)$—with t_0 the time of emergence from the last spectroscopy or state-selecting chamber—include the effects of one or more rf fields on the state of the atom prior to its decay to final states (f), whereas the initial elements of the density matrix σ of chapter 3 characterized atoms immediately after their creation in the target chamber. For a hydrogen atom in an oscillating field $E_0 \cos(\omega t + \delta)$ directed along the axis of quantization, electric dipole transitions can be induced between two, three, or four atomic levels (see figure 3.2) subject to the angular momentum selection rules: (a) $|\Delta L| = 1$, (b) $|\Delta J| = 0, 1$, (c) $|\Delta F| = 0, 1$ (but no transition between two $F = 0$ states), and (d) $\Delta M_F = 0$. The resulting elements $\rho_{ii'}(t_0)$ are then fairly complicated algebraic functions of field strength and frequency.

The initial statistical operator of the radiation field is the vacuum state projection operator

$$\rho^{(r)} = |0\rangle \langle 0|. \tag{3}$$

As described in the preceding chapter, creation and annihilation operators $(a_{k\sigma}^\dagger, a_{k\sigma})$ of the vector potential field, Eq. (5.4), respectively generate and destroy single-photon excitations of the electromagnetic field $|k\sigma\rangle$ with matrix elements given by Eqs. (5.5a) and (5.5b). The state is an eigenstate of (a) the field Hamiltonian $\mathcal{H}_r$ with eigenfrequency ω_k, (b) the field linear momentum operator with eigenvector k, and (c) the helicity operator (scalar product of linear momentum and intrinsic photon spin) with eigenvalue $\sigma = \pm 1$. That the photon, which is familiarly said to be a "spin-1 particle," has only two, rather than three, polarization components is a consequence of its zero rest mass[3]; more precisely, it is a helicity-1 particle. Photons of helicity ± 1 correspond respectively to LCP and RCP states with polarization directions specified by two complex-valued unit vectors $\hat{e}_{+1}(k)$ and $\hat{e}_{-1}(k)$ that satisfy the relations

$$\hat{e}_\sigma^*(k) \cdot \hat{e}_{\sigma'}(k) = \delta_{\sigma\sigma'}, \tag{4a}$$

$$\hat{e}_\sigma^*(k) \times \hat{e}_{\sigma'}(k) = i\sigma\hat{k}\delta_{\sigma\sigma'}. \tag{4b}$$

(A caret above a vector signifies a unit vector.)

The state of light can also be represented by a linearly polarized set of basis vectors $|\mathbf{k}\alpha\rangle$ $(\alpha = 1, 2)$ in which polarization directions are specified by two real-valued unit vectors $\hat{\mathbf{e}}_1(\mathbf{k})$ and $\hat{\mathbf{e}}_2(\mathbf{k})$ that satisfy orthonormalization relations comparable to Eqs. (4a, b):

$$\hat{\mathbf{e}}_\alpha(\mathbf{k}) \cdot \hat{\mathbf{e}}_{\alpha'}(\mathbf{k}) = \delta_{\alpha\alpha'}, \tag{5a}$$

$$\hat{\mathbf{e}}_\alpha(\mathbf{k}) \times \hat{\mathbf{e}}_{\alpha'}(\mathbf{k}) = \varepsilon_{\alpha\alpha'3}\mathbf{k}. \tag{5b}$$

Here ε_{ijk} is the completely antisymmetric tensor (Levi-Cività symbol) defined in chapter 2 (see note 7). Unit vectors for linear and circular polarizations can be expressed in terms of one another by means of the relation

$$\hat{\mathbf{e}}_\sigma(\mathbf{k}) = -\frac{\sigma}{\sqrt{2}}\left(\hat{\mathbf{e}}_1(\mathbf{k}) + i\sigma\hat{\mathbf{e}}_2(\mathbf{k})\right). \tag{6}$$

Correspondingly, the annihilation operators that act on linearly polarized basis states are related to the annihilation operators of helicity states in the following way:

$$a_{\mathbf{k}, 1} = -\frac{1}{\sqrt{2}}(a_{\mathbf{k}, +1} - a_{\mathbf{k}, -1}), \tag{7a}$$

$$a_{\mathbf{k}, 2} = -\frac{1}{\sqrt{2}}(a_{\mathbf{k}, +1} + a_{\mathbf{k}, -1}). \tag{7b}$$

Equations (7a) and (7b) follow from the requirement that the vacuum state be the same whether it is generated by annihilation of linearly polarized or circularly polarized single-photon states.

As in the previous chapter, a basis state of the combined atom–radiation field system will be written as $|\mu; \mathbf{k}\sigma\rangle$, and the statistical operator of the total system (before the interaction leading to radiative decay) is the direct product

$$\rho_i^{(a;r)} = \sum_{i,i'} |i; 0\rangle \rho_{ii'} \langle i'; 0|. \tag{8}$$

To lowest order in the field operators, the coupling of the atom and radiation field is effected by the interaction Hamiltonian $\mathcal{H}$ whose form is given in Eq. (5.2). Since the total Hamiltonian $\mathcal{H} = \mathcal{H}_a + \mathcal{H}_r + \mathcal{H}$ is independent of time, the statistical operator for the coupled atom–radiation field system evolves in time from Eq. (8) according to the transformation in Eq. (3.6):

$$\rho^{(a;r)}(t) = e^{-i\mathcal{H}t} \rho_i^{(a;r)} e^{i\mathcal{H}t} \tag{9}$$

(with the Hamiltonian again expressed in angular frequency units). Due to correlations between the individual atomic and radiation states, the operator $\rho^{(a;r)}(t)$ is in general no longer factorizable into a product of subsystem operators.

6.2 THE OPTICAL DETECTION FUNCTION

The act of detection may be viewed as the projection out of the totality of allowed final states just those states to which the detector (including state-selecting components like polarizers or filters) is sensitive. The probability that an observation is successful is determined in part by the physical laws governing the production of the states of interest and in part by the efficiency of the apparatus. In the experimental methodology combining optical detection and rf spectroscopy outlined in the preface, the detection process consists of counting photons of ideally well-defined energy and momentum. There may or may not be an additional selection of photons of specific polarization. A detection operator with these features can be expressed by

$$P^{(d)} = \sum_{f, \sigma, \sigma'} |f; \mathbf{k}\sigma\rangle \varepsilon_{\sigma\sigma'} \langle f; \mathbf{k}\sigma'|, \tag{10}$$

in which $\varepsilon_{\sigma\sigma'}$ is an element of the apparatus efficiency matrix in the helicity state representation. The element is $\varepsilon_{\sigma\sigma'}$ independent of the quantum numbers of the final atomic states $\{|f\rangle\}$ since these states are not observed. We will consider the construction and interpretation of the efficiency matrix in the next section, but for the present note that it must often be determined phenomenologically from a model of the behavior of the detecting apparatus.

We define next the optical detection function $\mathcal{W}$,

$$\mathcal{W} = \frac{2\pi}{\hbar} \text{Tr}\{P^{(d)}\mathcal{H}_I \rho_i^{(a;r)} \mathcal{H}_I^\dagger\} \eta(\omega_k), \tag{11}$$

which is a concise, representation-independent expression of the probability per unit time of observing the specified radiation states per unit solid angle subtended by the detector (that is, the Fermi golden rule in first-order perturbation theory).[4] The density of final radiation states $\eta(\omega_k)$ of energy $\hbar\omega_k$ and momentum $\hbar k$ per unit solid angle about k is given by Eq. (5.8b). Substitution into Eq. (11) of Eqs. (10), (5.2), and (8) leads to the explicit expression

$$\mathcal{W} = \left(\frac{e^2 \omega_k^3}{2\pi\hbar c^3}\right) \sum_{f, i, i'} \sum_{\sigma, \sigma'} \varepsilon_{\sigma\sigma'} \langle f | \mathbf{r} \cdot \hat{\mathbf{e}}_{\sigma'}^*(\mathbf{k}) | i \rangle \rho_{ii'} \langle i' | \mathbf{r} \cdot \hat{\mathbf{e}}_\sigma(\mathbf{k}) | f \rangle. \quad (12)$$

In arriving at Eq. (12), we have used the dipole approximation and the Heisenberg operator relation for velocity $d\mathbf{r}/dt = -i[\mathbf{r}, \mathcal{H}_a] = \mathbf{p}/m$ to transform momentum matrix elements into coordinate matrix elements as follows:

$$\langle f | \frac{\mathbf{p}}{m} \cdot \hat{\mathbf{e}}_\sigma^*(\mathbf{k}) e^{-i\mathbf{k}\cdot\mathbf{r}} | i \rangle \approx -i\omega_{if} \langle f | \mathbf{r} \cdot \hat{\mathbf{e}}_\sigma^*(\mathbf{k}) | i \rangle, \quad (13a)$$

$$\langle i | \frac{\mathbf{p}}{m} \cdot \hat{\mathbf{e}}_\sigma(\mathbf{k}) e^{i\mathbf{k}\cdot\mathbf{r}} | f \rangle \approx i\omega_{if} \langle i | \mathbf{r} \cdot \hat{\mathbf{e}}_\sigma(\mathbf{k}) | f \rangle, \quad (13b)$$

and set $\omega_{if} \approx \omega_k$, in keeping with the assumption of a quasi-monochromatic optical signal.

The angular distribution of the radiation is determined from the projection of the electron coordinate operator $\mathbf{r}$ onto the unit polarization vectors. To express this geometric dependence explicitly in terms of the laboratory frame of reference depicted in figure 6.1, we expand the vectors $\hat{\mathbf{e}}_\sigma(\mathbf{k})$ ($\sigma = \pm 1$) in linear superpositions of the cartesian unit vectors $\hat{\mathbf{x}}, \hat{\mathbf{y}}, \hat{\mathbf{z}}$:

$$\hat{\mathbf{e}}_\sigma(\mathbf{k}) = \sum_{\nu=0, \pm 1} D_{\nu\sigma}^1(z \to k) \hat{\mathbf{e}}_\nu \quad (14)$$

in which $\hat{\mathbf{e}}_0 = \hat{\mathbf{z}}$, $\hat{\mathbf{e}}_{\pm 1} = \mp (1/\sqrt{2})(\hat{\mathbf{x}} \pm i\hat{\mathbf{y}})$ [see Eq. (6)] constitute the three components of a spherical tensor of rank 1. The argument of the rotational matrix element $D_{\nu\sigma}^1(z \to k)$ is the set of Euler angles $(\phi_k, \theta_k, 0)$ for rotating the z-axis into the k-axis. The detection function then takes the form

$$\mathcal{W} = \mathcal{H} \sum_{f, i, i'} \sum_{\nu, \nu'} A_{\nu\nu'} \langle f | \mathbf{r} \cdot \hat{\mathbf{e}}_{\nu'}^* | i \rangle \rho_{ii'} \langle i' | \mathbf{r} \cdot \hat{\mathbf{e}}_\nu | f \rangle \quad (15)$$

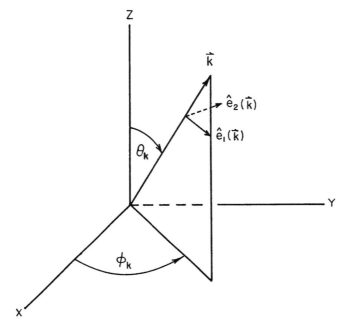

Figure 6.1 Disposition of the linear polarization vectors and propagation vector relative to the laboratory axis system. The triad of vectors satisfies $\hat{\mathbf{e}}_1(\mathbf{k}) \times \hat{\mathbf{e}}_2(\mathbf{k}) \cdot \hat{\mathbf{k}} = 1$.

where $\mathcal{H}$ is an unimportant constant factor, and the angular distribution function

$$
\begin{aligned}
A_{\nu\nu'} &= \sum_{\sigma,\,\sigma'} \varepsilon_{\sigma\sigma'} \left(D^1_{\nu'\sigma'}(z \to k) \right)^* D^1_{\nu\sigma}(z \to k) \\
&= \sum_{\sigma,\,\sigma'} D^1_{\nu\sigma}(z \to k) \varepsilon_{\sigma\sigma'} \left(D^1_{\sigma'\nu'}(z \to k) \right)^{-1}
\end{aligned}
\tag{16}
$$

contains the apparatus response and the angular dependence of the detected radiation.

The scalar products in Eq. (15) reduce to spherical tensors of rank 1:

$$
\mathbf{r} \cdot \hat{\mathbf{e}}_\nu^* = (-1)^{\nu+1} \sqrt{\frac{4\pi}{3}} r Y^1_{-\nu}(\theta,\,\phi),
\tag{17a}
$$

$$
\mathbf{r} \cdot \hat{\mathbf{e}}_\nu = (-1)^{\nu} \sqrt{\frac{4\pi}{3}} r Y^1_{\nu}(\theta,\,\phi),
\tag{17b}
$$

with Y_ν^1 a spherical harmonic. Each (hydrogenic) state vector in Eq. (15) comprises a linear superposition of electron orbital, electron spin, and nuclear spin states, whereas the operators in Eqs. (17a, b) act only in the Hilbert space of the orbital basis vectors. Thus, as in section 3.8, the matrix elements in Eq. (15) can be decomposed by means of the Racah algebra to yield the general expression

$$
\begin{aligned}
\langle nLSJIFM_F \mid T_Q^K(L, r) \mid & n'L'S'J'I'F'M_{F'} \rangle \\
= (-1)^{F+J+J'+L'-S-} & \, \delta_{SS'} \delta \quad , \\
& \times \sqrt{(2F'+1)(2J+1)(2J'+1)(2L+1)} \\
& \times W(JJ'FF'; KI) W(LL'JJ'; KS) \\
& \times \langle FM_F \mid F'KM_{F'}Q \rangle \langle nL \| T^K(L, r) \| n'L' \rangle,
\end{aligned}
\tag{18}
$$

in which $T_Q^K(L, r)$ is the Qth component of a tensor operator of rank K in the space of orbital wavefunctions, $\langle ab\alpha\beta | c -\gamma \rangle$ is a Clebsch–Gordan coefficient, and $W(abcd; ef)$ is a Racah coefficient. For the operators in Eqs. (17a) and (17b), the reduced matrix element $\langle nL \| T^1 \| n'L' \rangle$ is factorizable into an angular and radial portion,

$$
\langle nL \| T^1 \| n'L' \rangle = \langle L \| Y^1 \| L' \rangle \langle nL \| R \| n'L' \rangle,
\tag{19}
$$

where

$$
\langle L \| Y^1 \| L' \rangle = \sqrt{\frac{3}{4\pi} \frac{2L'+1}{2L+1}} \langle L0 \mid 1L'00 \rangle
\tag{20}
$$

and

$$
\langle nL \| R \| n'L' \rangle = \sqrt{\frac{4\pi}{3}} \int R_{nL} r^3 R_{n'L'} \, dr \equiv \sqrt{\frac{4\pi}{3}} R_{nL}^{n'L'}.
\tag{21}
$$

Here R_{nL} is the radial component of the hydrogen wavefunction.

When the preceding results are applied to the matrix elements in Eq. (15) (for which $K = 1$ and $Q = 0$), and the Clebsch–Gordan and Racah coefficients are converted to the more symmetrical 3-J and 6-J symbols, we obtain the optical detection function in a form convenient for direct evaluation:

$$
\begin{aligned}
\mathcal{W} = \mathcal{K} \sum_{f,i,i'} \sum_{\nu,\nu'} & (-1)^{J_i+J_{i'}-M_i-M_{i'}} A_{\nu\nu'} R^{n_iL_i}_{n_fL_f} R^{n_{i'}L_{i'}}_{n_fL_f} \rho_{ii'} \\
& \times (2F_f+1)(2J_f+1)(2L_f+1) \\
& \times \left[(2F_i+1)(2F_{i'}+1)(2J_i+1)(2J_{i'}+1)(2L_i+1)(2L_{i'}+1)\right]^{1/2} \\
& \times \begin{pmatrix} L_i & 1 & L_f \\ 0 & 0 & 0 \end{pmatrix} \begin{pmatrix} L_{i'} & 1 & L_f \\ 0 & 0 & 0 \end{pmatrix} \\
& \times \begin{pmatrix} F_f & 1 & F_i \\ M_f & \nu' & -M_i \end{pmatrix} \begin{pmatrix} F_f & 1 & F_{i'} \\ M_f & \nu & -M_{i'} \end{pmatrix} \\
& \times \begin{Bmatrix} J_i & J_f & 1 \\ F_f & F_i & \tfrac{1}{2} \end{Bmatrix} \begin{Bmatrix} J_{i'} & J_f & 1 \\ F_f & F_{i'} & \tfrac{1}{2} \end{Bmatrix} \\
& \times \begin{Bmatrix} L_i & L_f & 1 \\ J_f & J_i & \tfrac{1}{2} \end{Bmatrix} \begin{Bmatrix} L_{i'} & L_f & 1 \\ J_f & J_{i'} & \tfrac{1}{2} \end{Bmatrix}.
\end{aligned} \tag{22}
$$

Let us examine next the efficiency matrix ε.

6.3 THE EFFICIENCY MATRIX

The elements of the efficiency matrix e are dependent upon various geometric features of the detection process and on the efficiency of response of the apparatus to the incident photons. To keep the analysis simple but realistic, we will restrict the term "apparatus" to the primary optical elements affecting the intensity and polarization of the light, such as interference filters, polarizers, retarders, or lenses. A photon transmitted by these components and reaching a photomultiplier tube or other photon-counting device will be assumed detected.

Consider first the transmittance of a photon propagating along the normal to the surface of a filter-polarizer combination, which we shall refer to simply as a polarizer. Since only photons of a specified energy and polarization parallel to the transmission axis of the polarizer will be found on the output side, the action of the polarizer on the incident light can be represented by the projection operator[5]

$$P(\hat{e}) = |\hat{e}(\mathbf{k})\rangle\langle\hat{e}(\mathbf{k})|, \tag{23a}$$

in which $|\hat{e}(\mathbf{k})\rangle$ is the single-photon state of momentum $\hbar\mathbf{k}$ and unit polarization vector $\hat{e}(\mathbf{k})$. Upon expansion of $|\hat{e}(\mathbf{k})\rangle$ in a complete set of helicity basis states, Eq. (23a) becomes

$$\begin{aligned} P(\hat{e}) &= \sum_{\sigma,\sigma'} |\mathbf{k}\sigma\rangle\langle\mathbf{k}\sigma|\hat{e}(\mathbf{k})\rangle\langle\hat{e}(\mathbf{k})|\mathbf{k}\sigma'\rangle\langle\mathbf{k}\sigma'| \\ &\equiv \sum_{\sigma,\sigma'} |\mathbf{k}\sigma\rangle\varepsilon_{\sigma\sigma'}\langle\mathbf{k}\sigma'|, \end{aligned} \tag{23b}$$

which leads to the explicit form of the efficiency matrix elements,

$$\varepsilon_{\sigma\sigma'} = \langle\mathbf{k}\sigma \mid \hat{e}(\mathbf{k})\rangle\langle\hat{e}(\mathbf{k}) \mid \mathbf{k}\sigma'\rangle. \tag{24}$$

The diagonal elements give the probability that an incident photon of helicity σ is in a state to which the apparatus can respond, while the off-diagonal elements represent an apparatus-induced coherence between helicity states of orthogonal polarization.

A tacit assumption underlying the derivation of Eq. (24) is that the polarizer behaves ideally; that is, it passes with 100% efficiency photons polarized along $\hat{e}(\mathbf{k})$ and rejects with 100% efficiency orthogonally polarized photons. Under more realistic experimental conditions we will need to consider $\varepsilon_{\sigma\sigma'}$ to be phenomenological parameters whose values have to be obtained either empirically or from a model of the apparatus. Nevertheless, it is instructive to continue with the model of an ideal polarizer and to use Eq. (24) in the evaluation of ε. For high-quality commerical polarizers the transmittance of the favored polarization can be better than five orders of magnitude greater than that of the orthogonal component.[6]

Consider next the state $|\hat{e}(\mathbf{k})\rangle = |\mathbf{k}1(\delta)\rangle$, which is a linearly polarized basis state rotated by an angle δ about $\mathbf{k}$ so that $\hat{e}(\mathbf{k})$ is parallel to the linearly polarized basis vector $\hat{e}_1(\mathbf{k})$. The corresponding orthogonally polarized state is $|\mathbf{k}2(\delta)\rangle$. The state $|\mathbf{k}1(\delta)\rangle$ can be expanded in terms of helicity states:

$$|\mathbf{k}1(\delta)\rangle = \sum_{\sigma,\sigma'} |\mathbf{k}\sigma\rangle\langle\mathbf{k}\sigma' \mid \mathbf{k}\sigma'(\delta)\rangle\langle\mathbf{k}\sigma'(\delta) \mid \mathbf{k}1(\delta)\rangle, \tag{25}$$

in which $\langle \mathbf{k}\sigma \mid \mathbf{k}\sigma'(\delta) \rangle$ is a matrix element for the rotational transformation of a helicity basis state, and $\langle \mathbf{k}\sigma'(\delta) \mid \mathbf{k}1(\delta) \rangle$ is a matrix element for the transformation from a linear to a helicity representation. From angular momentum theory the first factor is readily shown to be

$$\langle \mathbf{k}\sigma \mid \mathbf{k}\sigma'(\delta) \rangle = D^1_{\sigma\sigma'}(\delta, 0, 0) = e^{-i\sigma\delta}\delta_{\sigma\sigma'}. \tag{26}$$

The second factor is invariant to rotations (in which case the specification of δ is unnecessary) and can be set equal to $\langle \mathbf{k}\sigma' \mid \mathbf{k}1 \rangle$. From Eqs. (7a) and (7b) we have

$$|\mathbf{k}\sigma\rangle = a^\dagger_{\mathbf{k}\sigma}|0\rangle = -\frac{\sigma}{\sqrt{2}}(a^\dagger_{\mathbf{k}1} + i\sigma a^\dagger_{\mathbf{k}2})|0\rangle = -\frac{\sigma}{\sqrt{2}}(|\mathbf{k}1\rangle + i\sigma|\mathbf{k}2\rangle), \tag{27}$$

from which follow immediately the scalar products

$$\langle \mathbf{k}\sigma \mid \mathbf{k}1 \rangle = -\frac{\sigma}{\sqrt{2}}, \tag{28a}$$

$$\langle \mathbf{k}\sigma \mid \mathbf{k}2 \rangle = \frac{i}{\sqrt{2}}. \tag{28b}$$

Substituting Eqs. (28a), (26), and (25) into Eq. (24), we find that

$$\varepsilon_{\sigma\sigma'} = \frac{\sigma\sigma'}{2}e^{-i(\sigma-\sigma')\delta} \tag{29a}$$

or equivalently

$$\varepsilon = \begin{pmatrix} \frac{1}{2} & -\frac{1}{2}e^{-2i\delta} \\ -\frac{1}{2}e^{2i\delta} & \frac{1}{2} \end{pmatrix}. \tag{29b}$$

Note that $\varepsilon^2 = \varepsilon$ and $\mathrm{Tr}\{\varepsilon^2\} = 1$ as is true for any density matrix characterizing a pure state.

For an incident beam of pure left or right circularly polarized photons, the statistical operator is respectively $\rho^{(r)}_{+1} = |\mathbf{k}, +1\rangle\langle\mathbf{k}, +1|$ or $\rho^{(r)}_{-1} = |\mathbf{k}, -1\rangle\langle\mathbf{k}, -1|$, and the probability of detection, $\mathrm{Tr}\{\rho^{(r)}_{\sigma=\pm1}P(\hat{\mathbf{e}})\} = \frac{1}{2}$, is the same for both polarizations and independent of the polarizing angle δ. For a superposition of the two circular polarizations, however, cross terms involving δ will affect the total transmission probability.

To express the efficiency matrix in a representation of linearly polarized basis states, we calculate the expectation value of $P(\hat{\mathbf{e}})$ in a state characterized by the statistical operator $\rho^{(r)} = |\mathbf{k}\alpha'\rangle\langle\mathbf{k}\alpha|$ (with $\alpha = 1, 2$). This leads to

$$\varepsilon_{\alpha\alpha'} = \text{Tr}\{|\mathbf{k}\alpha'\rangle\langle\mathbf{k}\alpha|P(\hat{\mathbf{e}})\}$$
$$= \sum_{\sigma, \sigma'} \langle\mathbf{k}\alpha' \mid \mathbf{k}\sigma\rangle\varepsilon_{\sigma\sigma'}\langle\mathbf{k}\sigma' \mid \mathbf{k}\alpha'\rangle \qquad (30a)$$

with

$$\varepsilon = \begin{pmatrix} \cos^2\delta & \frac{1}{2}\sin 2\delta \\ \frac{1}{2}\sin 2\delta & \cos^2\delta \end{pmatrix}, \qquad (30b)$$

where Eqs. (28a, b) and (29a) have been used to evaluate the transformation factors in Eq. (30a). An incident beam of linearly polarized photons characterized by the statistical operator $\rho_1^{(r)} = |\mathbf{k}1\rangle\langle\mathbf{k}1|$ or $\rho_2^{(r)} = |\mathbf{k}2\rangle\langle\mathbf{k}2|$ will therefore be detected with respective probabilities $\text{Tr}\{\rho_1^{(r)}P(\hat{\mathbf{e}})\} = \cos^2\delta$ and $\text{Tr}\{\rho_2^{(r)}P(\hat{\mathbf{e}})\} = \sin^2\delta$. The preceding results, interpreted as relative transmission intensities, are well known in classical optics as Malus's law.

For the ideal polarizer under discussion, the relation $\text{Tr}\{\varepsilon\} = 1$ holds independent of representation, and thus the total probability of finding a photon in either the transmitted beam or in the absorbed beam is unity. If the detecting apparatus is insensitive to the polarization of the incident light, the efficiency matrix $\bar{\varepsilon}$ is deduced either by averaging $\varepsilon(\delta)$ over all δ, or by summing the matrices for two orthogonal states of polarization:

$$\bar{\varepsilon}_{\sigma\sigma'} = \frac{1}{2\pi}\int_0^{2\pi} \varepsilon_{\sigma\sigma'}(\delta)\,d\delta = \varepsilon_{\sigma\sigma'}(0) + \varepsilon_{\sigma\sigma'}\left(\frac{\pi}{2}\right) = \tfrac{1}{2}\delta_{\sigma\sigma'}. \quad (31)$$

In that case the apparatus operator becomes

$$P(\hat{\mathbf{e}}) = \tfrac{1}{2}\sum_{\sigma=\pm 1} |\mathbf{k}\sigma\rangle\langle\mathbf{k}\sigma| = \tfrac{1}{2}\sum_{\alpha=1, 2} |\mathbf{k}\alpha\rangle\langle\mathbf{k}\alpha|, \qquad (32)$$

indicating that all incident photons will be detected. The matrix $\bar{\varepsilon}$ results from an incoherent mixture of two pure states; hence $\bar{\varepsilon}^2 \neq \bar{\varepsilon}$, and $\text{Tr}\{\bar{\varepsilon}^2\} < 1$.

We now relax the requirement of an ideal polarizer response and consider the case in which there can occur some transmission of photons polarized normal to $\hat{\mathbf{e}}(\mathbf{k})$. Let $\varepsilon_{(1)}$ be the probability of transmission of photons polarized parallel to $\hat{\mathbf{e}}(\mathbf{k})$ and $\varepsilon_{(2)}$ be the probability of leakage of the orthogonal component. Expressed in terms of a linearly polarized basis set with $\hat{e}_1(\mathbf{k})$ parallel to $\hat{e}(\mathbf{k})$, the apparatus operator becomes

$$P(\hat{e}) = \sum_\alpha |\mathbf{k}\alpha(\delta)\rangle \varepsilon_{(\alpha)} \langle \mathbf{k}\alpha(\delta)|. \tag{33}$$

For purposes of evaluating the angular distribution function $A_{\nu\nu'}$, Eq. (16), it is necessary to transform ε to a helicity-state representation. Using the previously derived transformation coefficients, we can write (with τ, τ' helicity indices)

$$\varepsilon_{\sigma\sigma'} = \sum_{\tau,\,\tau'} \sum_\alpha \langle \mathbf{k}\sigma \mid \mathbf{k}\tau(\delta)\rangle \langle \mathbf{k}\tau(\delta) \mid \mathbf{k}\alpha(\delta)\rangle \varepsilon_{(\alpha)}$$
$$\times \langle \mathbf{k}\alpha(\delta) \mid \mathbf{k}\tau'(\delta)\rangle \langle \mathbf{k}\tau'(\delta) \mid \mathbf{k}\sigma'\rangle, \tag{34a}$$

which leads to the matrix

$$\varepsilon = \begin{pmatrix} \frac{1}{2}(\varepsilon_{(1)} + \varepsilon_{(2)}) & -\frac{1}{2}(\varepsilon_{(1)} - \varepsilon_{(2)})e^{-2i\delta} \\ -\frac{1}{2}(\varepsilon_{(1)} - \varepsilon_{(2)})e^{2i\delta} & \frac{1}{2}(\varepsilon_{(1)} + \varepsilon_{(2)}) \end{pmatrix}. \tag{34b}$$

Without a more comprehensive theory of the polarizing mechanism, the parameters $\varepsilon_{(1)}$ and $\varepsilon_{(2)}$ must be determined empirically.

In concluding this section, we treat briefly the problem of a beam of light incident obliquely to the surface of the polarizer. A general quantum description of the process is made difficult by the fact that the state transmitted by the polarizer (with wavevector $\mathbf{k}$) is not simply expandable in terms of the states incident on the polarizer (with wavevector $\mathbf{k}' \neq \mathbf{k}$), and we turn instead to a classical description of the transmission process, using a simple model to obtain the angular dependence of ε. As an example of practical interest, consider an ideal linear polarizer of the dichroic film type in which the polarizing mechanism is based on selective absorption (by aligned molecules or crystals) of light polarized along the direction of alignment with negligible absorption of light polarized normal to this direction. The transmitted beam will then consist of that component of the initial beam whose polarization vector is orthogonal to the propagation direction and to the extinction axis $\hat{n}$ (that is, the orientation axis of the dichromophores).

As illustrated in Figure 6.2, we let $\mathbf{k}$ lie in the y–z plane and define the linear polarization vector $\hat{e}_2(\mathbf{k})$ to lie along the x-axis. The efficiency

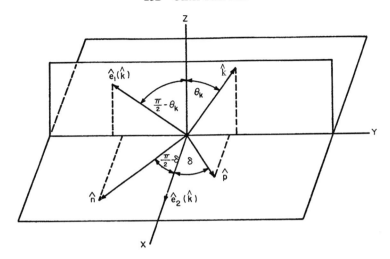

Figure 6.2 Oblique passage of light of propagation direction $\hat{\mathbf{k}}$ through a linear dichroic polarizer with polarizing axis $\hat{\mathbf{p}}$ and extinction axis $\hat{\mathbf{n}}$. Relative to the coordinate system shown, these directional vectors are $\hat{\mathbf{k}} = (0, \sin\theta_k, \cos\theta_k)$, $\hat{\mathbf{p}} = (\cos\delta, \sin\delta, 0)$, $\hat{\mathbf{n}} = (\sin\delta, -\cos\delta, 0)$. The polarization basis vectors are $\hat{\mathbf{e}}_1(\mathbf{k}) = (0, \cos\theta_k, \sin\theta_k)$ and $\hat{\mathbf{e}}_2(\mathbf{k}) = (1, 0, 0)$.

matrix for any other disposition of polarization vectors can be calculated by a rotation about the k-axis. The basis vectors $\hat{\mathbf{e}}_1(\mathbf{k})$ and $\hat{\mathbf{e}}_2(\mathbf{k})$ must now be resolved into components parallel to and perpendicular to $\hat{\mathbf{n}}$:

$$\hat{\mathbf{e}}_\alpha(\mathbf{k}) = \hat{\mathbf{e}}_\alpha^\parallel(\mathbf{k}) + \hat{\mathbf{e}}_\alpha^\perp(\mathbf{k}) = \left(\hat{\mathbf{e}}_\alpha(\mathbf{k}) \cdot \hat{\mathbf{n}}\right)\hat{\mathbf{n}} + \hat{\mathbf{n}} \times \left(\hat{\mathbf{e}}_\alpha(\mathbf{k}) \times \hat{\mathbf{n}}\right). \quad (35)$$

The same decomposition is then made for $\hat{\mathbf{e}}_\alpha^\perp(\mathbf{k})$ with respect to the **k**-direction:

$$\hat{\mathbf{e}}_\alpha^\perp(\mathbf{k}) = \hat{\mathbf{e}}_\alpha^{\perp\parallel}(\mathbf{k}) + \hat{\mathbf{e}}_\alpha^{\perp\perp}(\mathbf{k}) = \left(\hat{\mathbf{e}}_\alpha^\perp(\mathbf{k}) \cdot \hat{\mathbf{k}}\right)\hat{\mathbf{k}} + \hat{\mathbf{k}} \times \left(\hat{\mathbf{e}}_\alpha^\perp(\mathbf{k}) \times \hat{\mathbf{k}}\right). \quad (36)$$

Using the result of classical electromagnetic theory that the intensity of the transmitted light is proportional to the square of the electric vector, we can determine the elements of the efficiency matrix from the relation

$$\varepsilon_{\alpha\alpha'} = \eta\left|\hat{\mathbf{e}}_\alpha^{\perp\perp}\right|^2 \delta_{\alpha\alpha'} = \eta\left[1 - \left(\hat{\mathbf{e}}_\alpha(\mathbf{k}) \cdot \hat{\mathbf{n}}\right)^2\left(1 + (\hat{\mathbf{n}} \cdot \hat{\mathbf{k}})^2\right)\right]\delta_{\alpha\alpha'}, \quad (37a)$$

in which η is an instrumental constant. Expansion of the above scalar products in terms of the coordinates of figure 6.2 leads to matrix elements

$$\varepsilon_{11} = \eta\left[1 - \cos^2\theta\cos^2\delta(1 + \sin^2\theta\cos^2\delta)\right], \qquad (37\text{b})$$

$$\varepsilon_{12} = \eta\cos^2\delta(1 - \sin^2\theta\sin^2\delta), \qquad (37\text{c})$$

where δ is the angle between the polarizing axis and the x-axis, and θ is the angle between the transmitted wavevector and the z-axis. For the case of normal incidence ($\delta = 0$), Eqs. (37b, c) reduce to Malus's law [although note that δ is defined in this case with respect to $\hat{\mathbf{e}}_2(\mathbf{k})$, rather than $\hat{\mathbf{e}}_1(\mathbf{k})$ as before].

In many applications of a linear polarizer, the angles of interest are principally $\delta = 0$ and $\delta = \frac{1}{2}\pi$, in which cases Eqs. (37b, c) lead to

$$\varepsilon_{11}(\delta = 0) = \eta\sin^4\theta, \qquad \varepsilon_{22}(\delta = 0) = \eta, \qquad (38\text{a})$$

$$\varepsilon_{11}(\delta = \tfrac{1}{2}\pi) = \eta, \qquad \varepsilon_{22}(\delta = \tfrac{1}{2}\pi) = 0. \qquad (38\text{b})$$

Hence, within the range of validity of our model, obliquity effects should be small for $\theta \ll 1$. If the polarizer is set for $\delta = 0$ [so ideally only a beam with polarization $\hat{\mathbf{e}}_2(\mathbf{k})$ should pass], the extent of leakage for an angle as large as $\theta = 10°$ is only about 0.09%.

As a last point, it will be noted that $\mathrm{Tr}\{\varepsilon\}$ is not constant for angles of incidence differing from 0°. This is due to neglect of the polarization component $\hat{\mathbf{e}}_\alpha^{\perp\|}$ which is neither absorbed nor transmitted, but scattered.

6.4 THE OPTICAL SIGNAL

The optical detection function $\mathcal{W}$, defined by Eq. (11) and given explicitly in Eq. (15) (reproduced below),

$$\mathcal{W} = \mathcal{K}\sum_{f, i, i'}\sum_{\nu, \nu'}A_{\nu\nu'}\rho_{ii'}\langle f|\mathbf{r}\cdot\hat{\mathbf{e}}_{\nu'}^*|i\rangle\langle i'|\mathbf{r}\cdot\hat{\mathbf{e}}_\nu|f\rangle, \qquad (39)$$

is equivalent to the rate of production of specified radiation states per unit solid angle about the wavevector $\mathbf{k}$. Information relating to the production of the atomic states is contained in the initial conditions for $\rho_{ii'}$. The angular distribution and polarization of the emitted photons selected for detection are determined by the elements $A_{\nu\nu'}$; the probability of production of this radiation is governed by the dipole matrix elements.

In the optical detection of rf-induced transitions, photon counts are registered on two scalers switched synchronously with the rf field so that one scaler monitors light when the rf field is off and the other when the rf field is on. The actual optical signal corresponds not to $\mathcal{W}$, but to the function

$$\mathcal{S}(\omega, \mathrm{E}_0) = \mathcal{K}\big(\mathcal{W}(0) - \mathcal{W}(\omega, \mathrm{E}_0)\big), \tag{40}$$

in which $\mathcal{W}(\omega, \mathrm{E}_0)$ is the optical detection function with a radiofrequency field of frequency ω and amplitude E_0 activated, and $\mathcal{W}(0)$ is the corresponding function in the absence of such a field. We use the same symbol $\mathcal{K}$ to represent an arbitrary normalization constant that has no effect on the spectroscopic lineshapes. Substitution of $\mathcal{W}$ as given in Eq. (39) leads to the expression

$$\mathcal{S}(\omega, \mathrm{E}_0) = \mathcal{K} \sum_{f,\,[i,\,i']} \sum_{\nu,\,\nu'} A_{\nu\nu'}\big(\rho_{ii'}(0) - \rho_{ii'}(\omega, \mathrm{E}_0)\big)$$
$$\times \langle f|\mathbf{r}\cdot\hat{\mathbf{e}}_{\nu'}^*|i\rangle\langle i'|\mathbf{r}\cdot\hat{\mathbf{e}}_\nu|f\rangle, \tag{41}$$

where the bracketed indices signify summation over only those initial states affected by the rf field. This is a major simplification—and the reason for measuring $\mathcal{S}$ and not $\mathcal{W}$—since at a given frequency ω only a small number of initial states will be coupled; the remainder will be unaffected by the radiofrequency field and therefore will not contribute to the optical signal.

The detection procedure embodied in Eq. (40) permits one in principle to measure directly the light originating from each pair of coupled fine structure levels. If one of the coupled states does not radiate at the frequency being monitored (e.g., hydrogen $4F$ states cannot emit Balmer β light), then emission from a single fine structure level is detectable. An important advantage of this approach is that the major contribution to cascade radiation from higher electronic manifolds is limited to levels whose population can be altered by the applied field. Hence most of the cascade radiation has little effect on the measured lineshapes and the subsequent determination of level structure, lifetimes, or production cross sections.

Let us examine further the properties of $\mathcal{W}$ for several cases of experimental interest. The simplest example is that of field-free spontaneous emission from atoms produced at a thin foil target. For this case, the density matrix elements are given by Eq. (2) and lead to the optical detection function

$$\mathcal{W} = \mathcal{K} \sum_{f,i,i'} \sum_{\nu,\nu'} A_{\nu\nu'} e^{-(i\omega_{ii'} + \frac{1}{2}\gamma_{ii'})t} \rho^0_{ii'} \langle f | \mathbf{r} \cdot \hat{\mathbf{e}}^*_{\nu'} | i \rangle \langle i' | \mathbf{r} \cdot \hat{\mathbf{e}}_\nu | f \rangle. \quad (42)$$

The radiation is seen to be modulated at frequencies corresponding to the separation of pairs of levels for which the initial density matrix elements $\rho^0_{ii'}$ are nonvanishing. When observed along the beam, this temporal modulation is converted into a spatial modulation of the emitted light. Such zero-field oscillations or light beats provide an experimentally straightforward method for deducing atomic level separations.

The total optical detection function obtained by summing $\mathcal{W}$ over all the modes of the radiation field at the frequency ω_k displays a marked simplicity. Using the orthogonality properties of the rotation matrix and of the 3-J and 6-J symbols, together with the relation $\text{Tr}\{\varepsilon\} = 1$, yields the expression

$$\mathcal{W}_T = \int \mathcal{W}\, d\Omega = \mathcal{K} \sum_{i,f} \rho_{ii} \left| \left\langle n_i L_i \left\| \frac{\sqrt{4\pi}}{3} r Y^1 \right\| n_f L_f \right\rangle \right|^2 \quad (43)$$

with reduced matrix element (squared)

$$\left| \left\langle n_i L_i \left\| \frac{\sqrt{4\pi}}{3} r Y^1 \right\| n_f L_f \right\rangle \right|^2 = (2L_f + 1) \begin{pmatrix} L_i & 1 & L_f \\ 0 & 0 & 0 \end{pmatrix} (R^{n_i L_i}_{n_f L_f})^2. \quad (44)$$

Only the diagonal density matrix elements enter into the summation, and hence there is no temporal modulation of the total light output. Moreover, the dependence of $\mathcal{W}_T$ on the initial J, F, and M_F quantum numbers is confined to the elements of $\rho^{(a)}$; the optical transition matrix elements depend only on the n, L quantum numbers. This is a demonstration of the well-known principle that, within the electric dipole approximation, all states originating from the same orbital level have the same lifetime independent of the coupling to electron and nuclear spins. The factor $\sum_f |\langle n_i L_i \| \sqrt{4\pi/3} r Y^1 \| n_f L_f \rangle|^2$ is proportional to the total probability, $\gamma(i \to f)$, of transition from the initial level $|n_i L_i\rangle$ to the final levels $\{|n_f L_f\rangle\}$ of energy $\omega_f = \omega_i - \omega_k$. For example, the relative magnitudes of the factors $\gamma(i \to f)$ characterizing the hydrogen Balmer α and β lines are[7]

Balmer α: $\gamma(3S \to 2P) : \gamma(3P \to 2S) : \gamma(3D \to 2P) = 1 : 3.56 : 10.50$
Balmer β: $\gamma(4S \to 2P) : \gamma(4P \to 2S) : \gamma(4D \to 2P) = 1 : 3.75 : 8.00$

In practice, one does not ordinarily detect light radiated in all directions, but rather the small fraction emitted at a particular angle θ to the beam. If the detecting apparatus is insensitive to photon polarization, the elements of the efficiency matrix are simply $\varepsilon_{\sigma\sigma'} = \frac{1}{2}\delta_{\sigma\sigma'}$ and the angular distribution function [Eq. (16)] becomes diagonal. $\mathcal{W}$ then reduces to the expression

$$\mathcal{W} = \mathcal{K} \sum_{f, i, \nu} A_{\nu\nu} \rho_{ii} |\langle f | \mathbf{r} \cdot \hat{\mathbf{e}}_\nu | i \rangle|^2 \tag{45}$$

with

$$A_{\nu\nu} = 1 - |D^1_{\nu 0}(z \to k)|^2 \Rightarrow \begin{cases} A_{00} = \sin^2 \theta \\ A_{11} = A_{-1-1} = \frac{1}{2}(1 + \cos^2 \theta). \end{cases} \tag{46}$$

Note that only diagonal elements of the atomic density matrix again appear. In the absence of polarization selection, the optical signal exhibits *no* temporal modulation even if $\rho^{(a)}$ contains coherence terms. This unintuitive result can be established from the evaluation of $\mathcal{W}$ in Eq. (22). Set $\nu = \nu'$ (a consequence of the diagonal efficiency matrix) and apply the 3-J orthogonality relation (3.54a) in the sum over quantum numbers M_F and ν to find that the initial quantum numbers (F_i, M_i) and $(F_{i'}, M_{i'})$ must be equal, and therefore only diagonal terms ρ_{ii} appear in the final expression.

Strictly speaking, the foregoing results are valid for a point-source emitter and a detector of infinitesimal area set exactly at the angle θ to the beam (z-axis). In practice, of course, the detector subtends a finite solid angle and will admit photons with wavevectors lying within a certain range $\theta_2 \leq \theta \leq \theta_1$ and $\frac{1}{2}\pi - \phi_1 \leq \phi \leq \frac{1}{2}\pi + \phi_2$ of polar and azimuthal

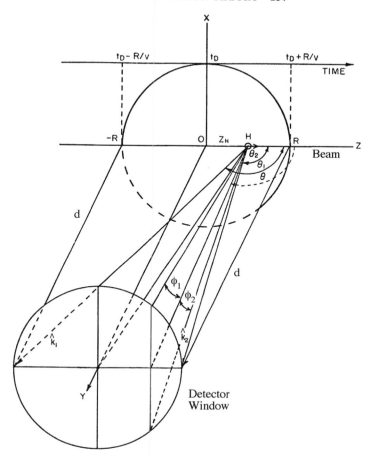

Figure 6.3 Geometry of the detection process for a finite-sized detector. $\hat{k}_1$ and $\hat{k}_2$ set limits on the propagation direction of detectable photons originating from an atom H at a point z_H in its trajectory across the detection chamber.

angles, as illustrated in figure 6.3. This problem is compounded by the fact that the source of emission is not stationary, but gives rise to an intensity that decreases exponentially across the detector window with a different decay rate for atoms in different orbital states. Moreover, the solid angle into which a photon can be emitted and detected depends on the location of the excited atom at the time of its decay.

Let t_D be the "detection time," that is, the length of time for an atom to travel at beam velocity v from the exit aperture of the target to a point $(z_0 = 0)$ on the line perpendicular to the detection window at its center.

From Eq. (45), the spontaneous emission from any atom decaying before or after t_D is then

$$\mathcal{W}(t_D, z) = \mathcal{K} \sum_{f, i, \nu} A_{\nu\nu} \rho_{ii}(t_D) |\langle f|\mathbf{r} \cdot \hat{\mathbf{e}}_\nu|i\rangle|^2 e^{-\gamma_i z/\upsilon} \tag{47}$$

and the integrated optical detection function (for a circular window of radius R) is

$$\overline{\mathcal{W}} = \frac{1}{2R} \int_{-R}^{R} dz \int_{\theta_2(z)}^{\theta_1(z)} \sin\theta\, d\theta \int_{\phi_2(z)}^{\phi_1(z)} d\phi \mathcal{W}(t_D, z). \tag{48a}$$

Equation (48a) can be expressed in the form

$$\overline{\mathcal{W}} = \mathcal{K} \sum_{f, i, \nu} J_{i;\nu} \rho_{ii}(t_D) |\langle f|\mathbf{r} \cdot \hat{\mathbf{e}}_\nu|i\rangle|^2, \tag{48b}$$

where, for the configuration shown in figure 6.3, the effect of finite detector size is contained in the integral

$$J_{i;\nu} = \frac{1}{R} \int_{-R}^{R} e^{-\gamma_i z/\upsilon} dz \int_{\theta_2(z)}^{\theta_1(z)} A_{\nu\nu}(\theta) \tan^{-1}\left(\frac{\sqrt{R^2 - z^2}}{d} \sin\theta\right) \sin\theta\, d\theta, \tag{49}$$

which in general must be evaluated numerically.

Fortunately, the refinements embodied in Eq. (48a) are ordinarily not necessary for interpreting optically detected rf lineshapes obtained in accelerator-based experiments where the detection geometry is well defined, the detector is insensitive to polarization, and the observed radiation is narrowly wavelength selected.

NOTES

1. H. Frauenfelder and R. M. Steffen, *Alpha-, Beta-, and Gamma-Ray Spectroscopy*, ed. by K. Siegbahn (North Holland, Amsterdam, 1965), pp. 997–1198.

2. M. P. Silverman and F. M. Pipkin, *J. Phys. B: Atom. Molec. Phys.* **7** (1974):730.

3. E. P. Wigner, *Rev. Mod. Phys.* **29** (1957):255.

4. For a reader interested in the rationale behind the selection of symbols, the traditional use of W in physics to represent a probability function derives ostensibly from the German word *Wahrscheinlichkeit* (probability).

5. An alternative discussion of the manipulation and detection of polarized light from the perspective of classical physical optics is given in M. P. Silverman, *Waves and Grains: Reflections on Light and Learning* (Princeton University Press, Princeton, NJ, 1998), chap. 9.

6. W. A. Shurcliff, *Polarized Light* (Harvard University Press, Cambridge, 1962).

7. H. Bethe and E. E. Salpeter, *Quantum Mechanics of One- and Two-Electron Atoms* (Springer, Berlin, 1957).

CHAPTER 7

State Selection and Lineshape Resolution

7.1 THE USE OF SEQUENTIAL FIELDS

A major impediment to the precise determination of atomic fine structure level separations by any spectroscopic method is the finite, and often broad, widths of the resulting lineshapes. For accelerator-based rf spectroscopy, in which a fast, monoenergetic beam is employed, the processes (discussed in chapter 1) of Doppler broadening (arising from a distribution of atomic velocities and hence interaction times) and collisional broadening (such as occurs in bulk gas) are usually insignificant. However, aside from instrumental effects, there are three intrinsic circumstances that can broaden rf lineshapes. One, discussed in earlier chapters, is the natural width due to spontaneous decay. Another is the overlapping of hyperfine transitions in an atom, like hyrogen, with a nuclear spin. The third is the possible overlap of transitions between different pairs of fine structure levels.

In the rf spectroscopy experiments pioneered by Lamb, a static magnetic field was used to separate the Zeeman components of each hydrogen fine structure level so that transitions of interest could be studied in a region of the microwave spectrum isolated from competing transitions. The magnetic field strength also served as the variable parameter to scan across a resonance. However, for the reasons outlined previously, there are disadvantages to the use of magnetic fields.

In this chapter we discuss the use of sequential oscillating fields (as illustrated in figure 7.1) to enhance the resolution of spectral lineshapes in the absence of magnetic fields. The frequency, strength, and relative orientation of these fields are judiciously chosen so as to remove from the beam, prior to its passage through the optical detection chamber, certain states whose induced transitions overlap the particular transition under study. Two conditions must hold for this method to be successful. First, one of the states giving rise to an overlapping transition must actually be "quenchable," that is, be coupled to another state of sufficiently

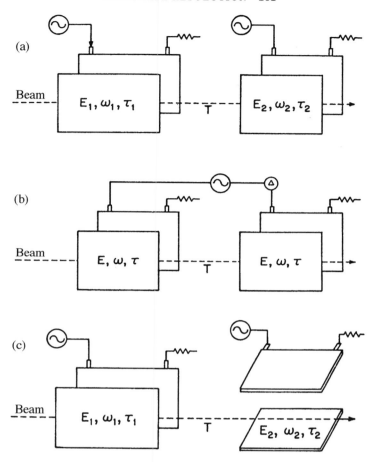

Figure 7.1 Configurations of sequential fields useful for spectroscopy: (a) Incoherent parallel fields, (b) coherent parallel fields with relative phase Δ, (c) incoherent nonparallel fields.

short lifetime that it will have decayed before reaching the optical detection chamber. A second condition, however, is that quenching of the undesired states take place without simultaneous removal of too many of the states of interest; in other words, that the quenching frequency lie outside the natural width of the resonance to be observed.

Once a single resonance profile has been isolated, additional resolution is attainable by using a spectroscopy field that is, itself, a sequential pair of coherently oscillating fields, a configuration first proposed by Ramsey[1] for magnetic resonance of stable molecular states.

The same idea can be fruitfully applied to the electric resonance of decaying states.[2] Such a configuration leads to narrowed lineshapes for two reasons. First, since a transition between states can occur in either one or the other of the two spatially separated field regions, there is a quantum interference of amplitudes that leads to "fringes" (in the frequency domain) across the resonance profile. Second, in the specific case of decaying states, the larger the separation between field regions, the more long lived are the surviving atoms that pass through the spectroscopy region to the optical detection chamber. The field configuration has, in effect, selected for spectroscopy a subpopulation of states of longer lifetime than the general population, with the consequence that the natural width of the resonance is narrowed.

In the following discussion, we will consider first the case of two parallel fields oscillating either incoherently or coherently (figures 7.1a, b), and then the case of two nonparallel fields (Figure 7.1c).

7.2 PARALLEL OSCILLATING FIELDS

An atom in a state defined by the initial wavefunction $\Psi(t_0)$ passes through a configuration of rf fields such that it experiences the field $\mathbf{E}_1(t) = E_1 \cos(\omega t + \delta)\hat{\mathbf{z}}$ for a time τ_1, then a region of field-free space for a time T, and last a second field $\mathbf{E}_2(t) = E_2 \cos(\omega' t + \delta')\hat{\mathbf{z}}$ for a time τ_2. Upon emerging after a total interaction time $t = \tau_1 + T + \tau_2$, the atom is in a state represented by $\Psi(t; t_0)$. We assume here that the axis of quantization of the beam is parallel to the two rf fields, and that the atomic states are coupled in pairs. Generalization to three or more coupled states is straightforward but algebraically tedious and of less immediate interest.

Using the results of previous chapters to obtain the atomic wavefunction at the time of passage through the first field, we can write

$$\Psi(t; t_0) = e^{-i\mathcal{H}_0\tau_1} K(\tau_1)\Psi(t_0)$$
$$= \begin{pmatrix} e^{-(i\omega_1 + \frac{1}{2}\gamma_1)t} & 0 \\ 0 & e^{-(i\omega_2 + \frac{1}{2}\gamma_2)t} \end{pmatrix}$$
$$\times \begin{pmatrix} K_{11}(\tau_1) & K_{12}(\tau_1)e^{-i\delta_1} \\ K_{21}(\tau_1)e^{i\delta_1} & K_{22}(\tau_1) \end{pmatrix} \Psi(t_0), \qquad (1)$$

in which the elements of the interaction matrix are given by

$$K_{11}(\tau_1) = e^{(\Gamma - i\Omega)\tau_1/2}\left[\cosh\nu\tau_1 - \left(\frac{(\Gamma - \delta\Gamma) - i(\Omega - \delta\Omega)}{2\nu}\right)\sinh\nu\tau_1\right], \quad (2a)$$

$$K_{12}(\tau_1) = -ie^{(\Gamma - i\Omega)\tau_1/2}\left(\frac{V}{\nu}\right)\sinh\nu\tau_1, \quad (2b)$$

$$K_{21}(\tau_1) = -ie^{-(\Gamma - i\Omega)\tau_1/2}\left(\frac{V}{\nu}\right)\sinh\nu\tau_1, \quad (2c)$$

$$K_{22}(\tau_1) = e^{-(\Gamma - i\Omega)\tau_1/2}\left[\cosh\nu\tau_1 + \left(\frac{(\Gamma - \delta\Gamma) - i(\Omega - \delta\Omega)}{2\nu}\right)\sinh\nu\tau_1\right], \quad (2d)$$

with

$$\nu = \sqrt{\tfrac{1}{4}[(\Gamma - \delta\Gamma) - i(\Omega - \delta\Omega)]^2 - V^2}. \quad (2e)$$

The phase in the off-diagonal elements is $\delta_1 = \omega t_0 + \delta$. As defined in earlier chapters, Ω is the detuning of the applied frequency ω from the Bohr frequency ω_0, Γ is one-half the decay rate difference, and V is the electric dipole matrix element. Note that the matrix K explicitly includes the first-order shift in resonance frequency ($\delta\Omega$) and decay rates ($\delta\Gamma$) arising from the interaction with the rf field. [See Eqs. (2.35) and (2.36) or (4.78b) and (4.78c).]

After leaving the first field, the wavefunction evolves in time over the interval T according to the operator $e^{-i\mathcal{H}_0 T}$. To account for passage through the second rf field, we merely repeat the application of Eq. (1) where, in place of $\Psi(t_0)$, is now inserted $\Psi(T + \tau_1; t_0)$. We replace δ_1 by $\delta_2 = \omega'(T + \tau_1) + \delta'$ and designate the new transition matrix as $K'(\tau_2)$. The final wavefunction then becomes

$$\Psi(t; t_0) = e^{-i\mathcal{H}_0\tau_2}K'(\tau_2)e^{-i\mathcal{H}_0(T+\tau_1)}K(\tau_1)\Psi(t_0). \quad (3)$$

Multiplying the interior factors of Eq. (3) and suitably rearranging the results leads to the form of a single-field interaction,

$$\Psi(t; t_0) = e^{-i\mathcal{H}_0 t}M\Psi(t_0), \quad (4)$$

where the interaction matrix M has elements

$$M_{11} = K'_{11}K_{11} + K'_{12}K_{21}e^{\Gamma(T+\tau_1)-i\Phi}, \tag{5a}$$

$$M_{12} = e^{-i\delta_1}\left[K'_{11}K_{12} + K'_{12}K_{22}e^{\Gamma(T+\tau_1)-i\Phi}\right], \tag{5b}$$

$$M_{21} = e^{i\delta_1}\left[K'_{22}K_{21} + K'_{21}K_{11}e^{-\Gamma(T+\tau_1)+i\Phi}\right], \tag{5c}$$

$$M_{22} = K'_{22}K_{22} + K'_{21}K_{12}e^{-\Gamma(T+\tau_1)+i\Phi}, \tag{5d}$$

with relative phase

$$\Phi = (\omega' - \omega_0)(T + \tau_1) + (\omega' - \omega)(t_0) + (\delta' - \delta). \tag{5e}$$

The form of the elements in Eqs. (5a–d) is amenable to a simple physical interpretation. Consider, for example, the element M_{11}. This is the amplitude for an atom initially in state 1 to remain in state 1 after passage through both fields. One possibility, represented by the term $K'_{11}K_{11}$, is for the atom to undergo no transition in either field. A second possibility, represented by $K'_{12}K_{21}$, is for the atom to undergo a transition to state 2 in the first field and then another transition back to state 1 in the second field. The sequence of events is revealed by the order of the indices read from right to left. Similarly, element M_{21} is the amplitude for an atom initially in state 1 to undergo a transition to state 2, and this may be achieved by undergoing the transition either in the first field (term $K'_{22}K_{21}$) or in the second field (term $K'_{21}K_{11}$).

Whether or not the two quantum pathways for each possible outcome lead to an observable interference effect depends on the relative phase $\Delta = (\delta' - \delta)$ of the two fields. If the two fields are driven by the same oscillator and are of the same frequency, we say that they are oscillating coherently; the phase difference Δ is then a well-defined, time-independent parameter of the experiment. For noncoherent fields, δ (or equivalently δ_1) and Δ are distributed quantities. To ascertain the occurrence and nature of any quantum interference effects, we must construct the density matrix of the system.

Consider first the case of incoherently oscillating fields. The density matrix characterizing the atomic beam is derived from the defining relation

$$\rho(t) = \langle \Psi(t)\Psi^\dagger(t)\rangle_{\delta_1, \Delta} = \langle e^{-i\mathcal{H}_0 t} M\rho(t_0)M^\dagger e^{i\mathcal{H}_0^\dagger t}\rangle_{\delta_1, \Delta}, \tag{6}$$

in which $\rho(t_0) = \Psi(t_0)\Psi^\dagger(t_0)$ is the initial density matrix and $\langle\ \rangle_{\delta_1, \Delta}$ signifies the average over all possible values of the phases δ_1 and Δ. It is assumed in what follows that the atoms are initially prepared in an incoherent mixture of states

$$\rho(t_0) = \begin{pmatrix} \alpha & 0 \\ 0 & \beta \end{pmatrix} \tag{7}$$

with $\text{Tr}\{\rho(t_0)\} = \alpha + \beta = 1$. Substituting Eq. (7) into Eq. (6) and averaging both δ_1 and Δ over the range of angles from 0 to 2π yields

$$\rho_{11}(t) = e^{-\gamma_1 t}\big[\alpha\{|K'_{11}K_{11}|^2 + |K'_{12}K_{21}|^2 e^{2\Gamma(T+\tau_1)}\}$$
$$+ \beta\{|K'_{11}K_{12}|^2 + |K'_{12}K_{22}|^2 e^{2\Gamma(T+\tau_1)}\}\big], \tag{8a}$$

$$\rho_{22}(t) = e^{-\gamma_2 t}\big[\alpha\{|K'_{22}K_{11}|^2 + |K'_{21}K_{11}|^2 e^{-2\Gamma(T+\tau_1)}\}$$
$$+ \beta\{|K'_{22}K_{22}|^2 + |K'_{21}K_{12}|^2 e^{-2\Gamma(T+\tau_1)}\}\big]. \tag{8b}$$

There are no off-diagonal elements. Moreover, Eqs. (8a, b) constitute a sum of the probabilities for the atom to emerge in state 1 and state 2, respectively; there is no interference between different quantum pathways.

In the case of coherently oscillating fields, the frequencies are the same, $\omega = \omega'$, and the phase average is performed only over δ_1. For the assumed initial condition of Eq. (7), this again leads to a diagonal density matrix, but the individual elements

$$\rho_{11}(t) = e^{-\gamma_1 t}\big[\alpha|K'_{11}K_{11} + K'_{12}K_{21}e^{(\Gamma-i\Omega)(T+\tau_1)}e^{-i\Delta}|^2$$
$$+ \beta|K'_{11}K_{12} + K'_{12}K_{22}e^{(\Gamma-i\Omega)(T+\tau_1)}e^{-i\Delta}|^2\big], \tag{9a}$$

$$\rho_{22}(t) = e^{-\gamma_2 t}\big[\alpha|K'_{22}K_{21} + K'_{21}K_{11}e^{-(\Gamma-i\Omega)(T+\tau_1)}e^{+i\Delta}|^2$$
$$+ \beta|K'_{22}K_{22} + K'_{21}K_{12}e^{-(\Gamma-i\Omega)(T+\tau_1)}e^{+i\Delta}|^2\big], \tag{9b}$$

contain cross-terms that represent the interference of quantum pathways for effecting each of the four possible processes: $1 \rightarrow 1$, $1 \rightarrow 2$, $2 \rightarrow 1$, $2 \rightarrow 2$.

If we set $\tau_1 = \tau_2 = \tau$ and $E_1 = E_2$, which are the usual experimental conditions for application of the separated oscillating-field method, and assume that the beam has been prepared so that $\alpha = 1$ and $\beta = 0$, the elements (9a, b) of the density matrix reduce to

$$\rho_{11}(t) = e^{-\gamma_1 t}\big[|K_{11}^2|^2 + |K_{12}K_{21}|^2 e^{2\Gamma(T+\tau)}$$
$$+ 2e^{\Gamma(T+\tau)}\operatorname{Re}\{K_{11}^2 K_{12}^* K_{21}^* e^{i\Omega(T+\tau)+i\Delta}\}\big], \tag{10a}$$

$$\rho_{22}(t) = e^{-\gamma_2 t}\big[|K_{22}K_{21}|^2 + |K_{21}K_{11}|^2 e^{-2\Gamma(T+\tau)}$$
$$+ 2e^{-\Gamma(T+\tau)}\operatorname{Re}\{K_{22}|K_{21}|^2 K_{11}^* e^{-i\Omega(T+\tau)-i\Delta}\}\big]. \tag{10b}$$

Of primary interest in each element is the interference term which is a function of Δ and thus absent in the density matrix elements (7a, b) of incoherent fields. In accord with the heuristic argument given earlier based on the uncertainty principle, note that the exponential phase factor of the interference term becomes increasingly oscillatory as the field-free interaction time T gets larger. At resonance ($\Omega = 0$) however, the interference term is independent of T, in which case the separation between interaction regions has no effect but to diminish the overall intensity if the states are unstable.

Figure 7.2 shows the effect of different values of Δ on ρ_{11} near the resonance frequency for transitions between hydrogen states $|1\rangle = |3^2 S_{1/2}\rangle$ and $|2\rangle = |3^2 P_{1/2}\rangle$. We have neglected here the ac Stark effect. For $\Delta = \pm \frac{1}{2}\pi$ the minimum of the resonance profile shifts to a frequency lower and higher, respectively, than ω_0, but the magnitude of ρ_{11} is the same. For $\Delta = 0, \pi$ the magnitude of ρ_{11} at resonance is smaller and greater, respectively, than the corresponding magnitude of the phase-averaged value $\langle\rho_{11}\rangle_\Delta$, but the resonance frequency is unshifted. This behavior is analogous to that pointed out in chapter 3 concerning the effect of the phase δ on the single-field two-state lineshape.

The case of exact resonance ($\Omega = 0$) between two states of the same lifetime ($\Gamma = 0$) provides a good example to understand better the physical effect of the relative phase Δ. Under these circumstances, $\nu = iV$, the interaction matrix elements $K_{11} = K_{22}$ are real valued, $K_{12} = K_{21}$ are purely imaginary, and the relative populations of states 1 and 2 become

$$\rho_{11} = e^{-\gamma_1 t}(\cos^4 V\tau + \sin^4 V\tau - 2\cos^2(V\tau)\sin^2(V\tau)\cos\Delta), \tag{11a}$$
$$\rho_{22} = 2e^{-\gamma_2 t}\cos^2 V\tau \sin^2 V\tau(1 + \cos\Delta). \tag{11b}$$

Note that in the absence of decay expressions (11a) and (11b) sum to unity irrespective of Δ; the relative phase cannot, of course, affect the

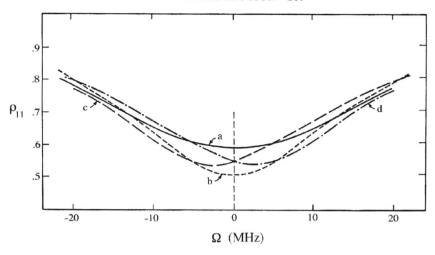

Figure 7.2 Effect of relative phase Δ on ρ_{11} near the resonance frequency $\Omega = 0$. $\Delta =$ (a) 0, (b) π, (c) $+\pi/2$, (d) $-\pi/2$. Parameters of the calculation are interaction time $\tau = 60.3$ ns, field-free transit time $T = 0$, decay rates $\gamma_1 = 6.25 \times 10^6$ s^{-1}, $\gamma_2 = 172.4 \times 10^6$ s^{-1}, and coupling strength $V = 16$ MHz.

conservation of probability. Choosing the field strength and interaction time so that $V\tau = \frac{1}{4}\pi$ reduces Eqs. (11a, b) to the expressions

$$P_{11} = \tfrac{1}{2}e^{-\gamma_1 t}(1 - \cos\Delta), \tag{11c}$$

$$P_{22} = \tfrac{1}{2}e^{-\gamma_2 t}(1 + \cos\Delta), \tag{11d}$$

which vary periodically with Δ with 100% contrast. If the two coherently oscillating fields are in phase ($\Delta = 0$), then the atom, initially in state 1, emerges in state 2 with 100% probability. On the other hand, if the two fields are 180° out of phase ($\Delta = \pi$), the emerging atom will be found in the original state 1 with 100% probability.

When the frequency is not exactly at resonance or the states are not stable, the interference term still diminishes the component of state 1 in the wavefunction for fields oscillating in phase and augments this component for fields oscillating out of phase. However, the contrast in these cases may be too low for different choices of Δ to be noticeable in the direct variation of ρ_{11} or ρ_{22} with ω. The properties illustrated in figure 7.2 then serve to practical advantage in spectroscopy. By measuring

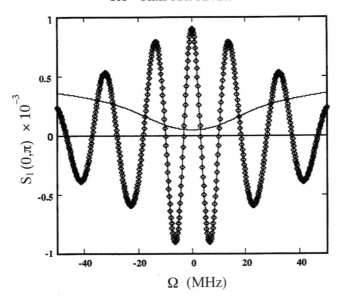

Figure 7.3 Isolation of quantum interference in the relative population ρ_{11} of a long-lived state ($3^2S_{1/2}$) coupled to a short-lived state ($3^2P_{1/2}$) by two coherently oscillating fields. The fringe pattern is the difference in ρ_{11} (as a function of detuning frequency) for relative phase $\Delta = 0$ and π. This pattern is not visible in the frequency variation of ρ_{11} itself (superposed curve). Theoretical parameters are $V = 6$ MHz, $\tau = 40$ ns, $T = 50$ ns; decay constants are the same as for figure 7.2.

the difference in occupation probability for two values of Δ,

$$S_i(\Delta_1, \Delta_2) = \rho_{ii}(\Delta_1) - \rho_{ii}(\Delta_2) \tag{12}$$

where $|\Delta_1 - \Delta_2| = \pi$, one can eliminate from the lineshape the dominant phase-independent contributions and thereby accentuate the small differences engendered by the interference term. The procedure defined by Eq. (12) leads to the theoretical signals

$$S_1(0, \pi) = 4e^{-\gamma_1 t}e^{\Gamma(T+\tau)} \, \text{Re}\{K_{11}^2 K_{12}^* K_{21}^* e^{i\Omega(T+\tau)}\}, \tag{13a}$$

$$S_2(0, \pi) = 4e^{-\gamma_2 t}e^{-\Gamma(T+\tau)} \, \text{Re}\{K_{22}|K_{21}|^2 K_{11}^* e^{-i\Omega(T+\tau)}\}, \tag{13b}$$

$$S_1(-\tfrac{1}{2}\pi, \tfrac{1}{2}\pi) = 4e^{-\gamma_1 t}e^{\Gamma(T+\tau)} \, \text{Im}\{K_{11}^2 K_{12}^* K_{21}^* e^{i\Omega(T+\tau)}\}, \tag{13c}$$

$$S_2(-\tfrac{1}{2}\pi, \tfrac{1}{2}\pi) = -4e^{-\gamma_2 t}e^{-\Gamma(T+\tau)} \, \text{Im}\{K_{22}|K_{21}|^2 K_{11}^* e^{-i\Omega(T+\tau)}\}. \tag{13d}$$

Figure 7.3 compares the variation with frequency of $S_1(0, \pi)$ and $\rho_{11}(\Delta = 0)$ in the case of two coupled unstable states (hydrogen 3S and

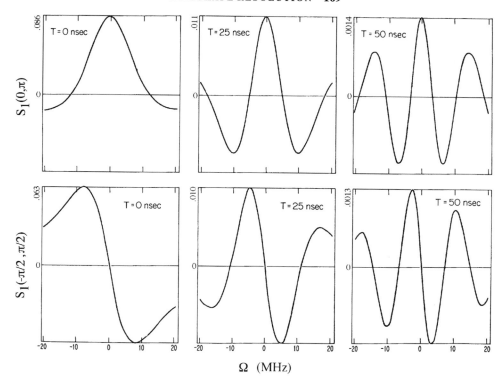

Figure 7.4 Narrowing of the fringe patterns of $S_1(0, \pi)$ and $S_1(-\frac{1}{2}\pi, \frac{1}{2}\pi)$ with increasing field-free transit time T for a fixed interaction time of $t = 60$ ns. The decay rates and coupling strength are the same as for figure 7.2.

3P states). The scale of the figure shows that the interference term is approximately one hundred times smaller than ρ_{11} in the vicinity of resonance, and therefore the visibility of the interference pattern is greatly enhanced in the difference signal $S_1(0, \pi)$. Because the central fringe of the interference lineshape is much narrower than the width of ρ_{11}, the difference signal (12) allows one to locate the experimental peak of a resonance with greater precision than by a single-field experiment. As already pointed out, the longer the transit time T between interaction regions, the narrower the fringes become. This is illustrated in figure 7.4 which also shows the effects of switching the relative phase Δ between $\pm\frac{1}{2}\pi$.

In general, the signal $S_i(0, \pi)$ $(i = 1, 2)$ as a function of frequency Ω gives an absorption-like curve, whereas the signal $S_i(-\frac{1}{2}\pi, \frac{1}{2}\pi)$ results in a dispersion-like curve. In the absence of the counter-rotating component of the oscillating field, $S_i(0, \pi)$ is perfectly symmetric about the resonance frequency $(\Omega = 0)$, and $S_i(-\frac{1}{2}\pi, \frac{1}{2}\pi)$ is antisymmetric. The latter signal is also spectroscopically useful, for there are occasions where it is more advantageous to determine the zero-crossing point in the region of steepest slope of a resonance line than to locate the maximum or minimum point in a region of near-zero slope. A lineshape of the dispersion type is particularly sensitive to small shifts in the resonance frequency. In this regard it should be noted that the ac Stark shift derived previously for a single oscillating field [Eq. (2.35)] does not necessarily apply to a pair of coherently oscillating fields. For such a configuration one can calculate the resonance displacement by solving the equation $(d/d\Omega)\, S(\Delta_1, \Delta_2) = 0$ for Ω; the result, which depends on the beam velocity, is a complicated function of the times τ and T.

7.3 NONPARALLEL OSCILLATING FIELDS

In the rf field configuration of figure 7.1c the direction of the second field $\mathbf{E}_2(t)$ is specified by the polar and azimuthal angles θ, ϕ with respect to a coordinate system (x, y, z) in which the first field $\mathbf{E}_1(t)$ is parallel to z. After interaction with the first field and passage through field-free space, the atom is in the state

$$
\begin{aligned}
\Psi(T + \tau_1; t_0) = &\sum_M c_{\alpha F_1 M}(\tau_1; t_0) e^{-iX_\alpha T} |\alpha F_1 M\rangle_z \\
&+ \sum_N c_{\beta F_2 N}(\tau_1; t_0) e^{-iX_\beta T} |\beta F_2 N\rangle_z,
\end{aligned} \qquad (14)
$$

in which the labels α and β denote the residual quantum numbers of two hyperfine states of opposite parity, $X_\mu = \omega_\mu - i\frac{1}{2}\gamma_\mu$ $(\mu = \alpha, \beta)$ symbolizes the (complex) energy eigenvalues of the uncoupled atom, and the amplitudes $c_{\alpha F_1 M}(\tau_1; t_0)$ and $c_{\beta F_2 N}(\tau_1; t_0)$ can be calculated by the methods of chapter 3. The subscript z on the basis vectors indicates explicitly that the states are defined with respect to field $\mathbf{E}_1$ as the quantization axis.

To determine the effect of the second rf field, we express the basis states $|\rangle_z$ in terms of the states $|\rangle_{z'}$ where $\mathbf{E}_2$ is parallel to z'. Unlike the analysis of the preceding section, it is now necessary to consider all Zeeman states within each coupled multiplet. If the coordinate system (x', y', z') is generated from the original system (x, y, z) by a rotational transformation R of ζ radians about the unit vector $\hat{\mathbf{n}}$, then the eigenvectors $|\rangle_{z'}$ of the projection $F_{z'}$ of the total angular momentum operator $\mathbf{F}$ are

$$|\rangle_{z'} = D(R)|\rangle_z, \tag{15}$$

where $D(R) = e^{-i\hat{\mathbf{n}}\cdot\mathbf{F}\zeta}$ is the rotation operator. From the inverse transformation governed by the rotational operator $D^{-1}(R) = D(R^{-1})$, we can express the states $|\rangle_z$ as superpositions of the states $|\rangle_{z'}$:

$$|\mu FM\rangle_z = \sum_{M'} D^F_{M'M}(R^{-1})|\mu FM'\rangle_{z'}. \tag{16}$$

The argument of the operator $D(R)$ is the set of Euler angles $(\phi, \theta, 0)$ of the rotation $(x, y, z) \to (x', y', z')$. In terms of the new basis states, the state vector (14) becomes

$$\begin{aligned}
\Psi(T + \tau_1; t_0) = &\sum_{M'} d_{\alpha F_1 M'} e^{-iX_\alpha T}|\alpha F_1 M'\rangle_{z'} \\
&+ \sum_{N'} d_{\beta F_2 N'} e^{-iX_\beta T}|\beta F_2 N'\rangle_{z'},
\end{aligned} \tag{17}$$

where

$$d_{\mu FM'}(\tau_1; t_0) = \sum_M D^F_{M'M}(R^{-1})c_{\mu FM}(\tau_1; t_0) \tag{18}$$

constitutes the new set of initial conditions for the interaction of the atom with the field $\mathbf{E}_2(t)$. This interaction can be analyzed by the methods of chapter 3 since all states are again either uncoupled or coupled in groups of two, three, or four. Last, an inverse rotational transformation yields the amplitudes $c_{\alpha F_1 M}(t)$ and $c_{\beta F_2 N}(t)$ for states defined with respect to the original quantization axis. Proper phase averaging of bilinear products of the amplitudes leads to the elements of the density matrix. The procedure outlined here is equivalent to that employed in section 3.3 of chapter 3 to express the density matrix in terms of a different quantization axis.

One application of the foregoing procedure is in the use of two perpendicular rf fields as a hyperfine state selector for the hydrogenic state $|^2S_{1/2}00\rangle$ when the initial beam includes all magnetic substates of the $S(F = 1)$ and $S(F = 0)$ hyperfine multiplets. Upon passage through a first rf field tuned to the resonance frequency for $^2S_{1/2}(F = 1) \to$ $^2P_{1/2}(F = 1)$ transitions, the $|^2S_{1/2}1, \pm1\rangle_z$ states will be quenched, and the emerging beam will contain a mixture of $|^2S_{1/2}00\rangle_z$ and $|^2S_{1/2}10\rangle_z$ states. (We disregard the rapidly decaying P states.) Since the matrix element $_z\langle^2S_{1.2}10|z|^2P_{1/2}10\rangle_z = 0$, the $^2S_{1/2}(F = 1, M_F = 0)$ atoms cannot be eliminated by the first field. The beam is then passed through a second field of the same frequency, but rotated 90° with respect to the first field. With respect to this new axis of quantization (the x-axis), the state $|^2S_{1/2}10\rangle_z$ takes the form

$$|^2S_{1/2}10\rangle_z = \sum_{M'} D^1_{M'0}(0, -\tfrac{1}{2}\pi, 0)\, |^2S_{1/2}1M'\rangle_x$$

$$= \frac{1}{\sqrt{2}}(|^2S_{1/2}11\rangle_x - |^2S_{1/2}1\bar{1}\rangle_x). \tag{19}$$

These states, coupled to $|^2P_{1/2}11\rangle_x$ and $|^2P_{1/2}1\bar{1}\rangle_x$ by the second field, are removed from the beam which, shortly after emerging, consists solely of atoms in the desired state $|^2S_{1/2}00\rangle$. No quantization axis need be indicated in this case since the state is spherically symmetric.

NOTES

1. N. F. Ramsey, *Molecular Beams* (Oxford University Press, London, 1956), pp. 124–134.

2. M. P. Silverman, *Optical Electric Resonance Investigation of a Fast Hydrogen Beam* (Ph.D. Thesis, Harvard University, 1973).

CHAPTER 8

Elements of Experimental Design and Application

8.1 GENERAL DESCRIPTION

Technology advances so rapidly that what may pass for the state of the art at the time of writing is often outmoded by the date of publication. It is therefore not the objective of this chapter to survey developments in apparatus design or numerous spectroscopic studies in which accelerators have been employed. Rather, it is instructive to consider basic, general aspects, both practical and conceptual, of how one actually makes a fast beam of neutral atoms and ensures that subsequently induced atomic transitions occur under the conditions presupposed by theory. To this end what may serve as the best example for discussion is a simple, rather than sophisticated, apparatus design; namely, the home-made fast-beam electric resonance spectrometer (Figure 8.1) with which the SABER method of probing atoms was first successfully implemented.

As shown in the figure the spectrometer comprises five principal subsystems:

  1. Ion production (with an inductively coupled radiofrequency ion source) and extraction

  2. Ion acceleration and focusing (with a cylindrical unipotential electrostatic gap lens)

  3. Excited atom production (through charge-exchange collisions within a gas target [as shown], or with a carbon foil target [not shown])

  4. Radiofrequency state selection and spectroscopy (by means of one or more parallel-plate capacitor assemblies)

  5. Optical detection and signal processing (with a photomultiplier tube and associated fast electronics)

Before each of these subsystems is discussed individually, two generic features of an atomic-beam apparatus should be noted.

First, common to virtually all atomic-beam experiments, is the need to maintain a high vacuum (in the present case, less than $\sim 10^{-6}$ mm Hg or $\sim 10^{-4}$ Pa) if atoms are to propagate collision-free down the beam line,

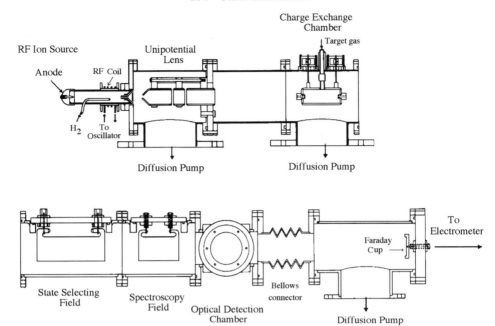

Figure 8.1 Cross section through the fast-beam electric resonance spectrometer showing all component systems.

and electric arcing from high-voltage components is to be avoided. In a fast-beam experiment there are two large sources of gas: the ion source and the gas target. To handle this load, the apparatus of figure 8.1 was evacuated by three oil diffusion pumps (each with a pumping capacity of about 700 liters per second) backed by conventional rotary pumps. To keep oil from entering the experimental chambers and depositing on metal surfaces (thereby forming insulating layers that give rise to stray electric fields), the rotary pumps were fitted with molecular-sieve traps and the diffusion pumps with thermoelectric or liquid-nitrogen-cooled baffles. For a comprehensive discussion of vacuum technology, the reader can consult a number of excellent reference books.[1]

Second, the beam line is surrounded by three mutually orthogonal pairs of coils termed Helmholtz coils (not shown), which serve to cancel the magnetic field of the earth to within less than 10 mG inside the atom–rf field interaction region. Without such cancellation, the earth's magnetic field would give rise to (at least) two uncontrolled perturbations: (a) Zeeman splitting of the atomic fine or hyperfine states, and (b) a "motional" Stark effect. The latter effect[2] arises from the relativistic transformation of electric and magnetic fields. A charged particle moving with velocity $\mathbf{v}$ in the presence of a magnetic field $\mathbf{B}$ experiences in its rest frame an electric field $\mathbf{E} = \left(1/\sqrt{1-(v/c)^2}\right)\mathbf{v} \times \mathbf{B}$ that can split or broaden resonance lineshapes.

Let us now consider the individual subsystems.

8.2 ION PRODUCTION AND EXTRACTION

The first proton accelerator was constructed by Cockroft and Walton[3] in 1932, and since that time sources for the production of ions—especially of hydrogen, deuterium, and helium—have become an indispensable part of accelerating machines for purposes as diverse as nuclear reaction studies, atomic spectroscopy, and ion microscopy. Since the hydrogen atom is the principal system of interest in this book, the present discussion focuses on the production of protons. Nevertheless, the type of source known as a radiofrequency ion source, shown in greater detail in figure 8.2, has also been used to produce deuterons, He^{3+}, He^{4+}, and singly charged ions of argon and xenon, among other species. The rf field accelerates free electrons inside the bottle to energies sufficient to dissociate or ionize the gas.

Some eighty years ago, the noted optical physicist R. W. Wood[4] discovered that a hydrogen discharge in a long tube of clean Pyrex glass with high-voltage electrodes at the ends exhibited an almost pure Balmer spectrum near the midsection. At the ends near the metal electrodes, the discharge exhibited the spectrum of molecular hydrogen, indicating the almost complete occurrence of recombination. Following Wood's work, J. J. and G. P. Thomson[5] studied electrical conduction through ionized gases in discharge vessels containing no metal, the discharge being supported by an external induction coil through which the Pyrex tube was inserted. Here again, the most intense part of the discharge exhibited a virtually pure Balmer spectrum. Pyrex, as was subsequently learned, has the lowest recombination coefficient of any known substance, a value of 2×10^{-5} on a scale where the coefficient of nearly all metals is close to unity.

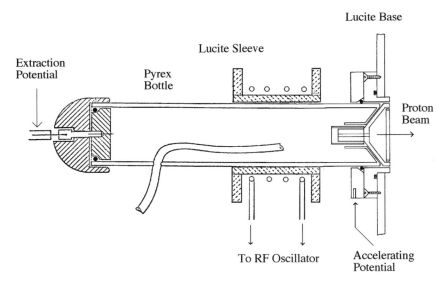

Figure 8.2 Detailed anatomy of the rf ion source.

Ideally, a hydrogen ion source employed with an accelerator should be stable and long lived, yield a large beam current for low gas consumption with protons as the major component, consume low power, and be simple and compact in its construction. An electrodeless rf discharge source, such as illustrated in figure 8.2 (adapted from a model by Moak, Reese, and Good [MRG][6]), meets these criteria quite well.

Hydrogen gas is slowly admitted at a rate of 1–2 cm^3/hr through a long coiled Pyrex tube which serves to increase the breakdown path to the grounded copper feedline (not shown) from the supply tank. It is quite important that the feed system be free of leaks, for the quality of the discharge is severely impaired by extraneous gases. The admission of air, for example, changes the color of the discharge from a rich crimson (the Balmer red Hα and blue Hβ lines) to a sallow pink (from molecular nitrogen).

Figure 8.3 illustrates two commonly employed types of rf coupling by means of which a plasma is formed and sustained within the tube. Capacitive coupling, which was employed by MRG, gives rise to a "linear" discharge, so called because the electric field is parallel to the tube axis. Under comparable conditions of rf power and source pressure, the "annular" discharge engendered by inductive coupling has been said to

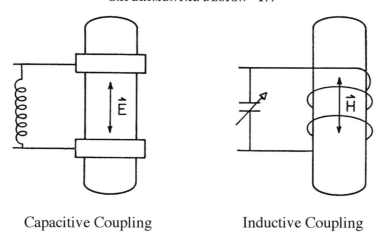

Capacitive Coupling   Inductive Coupling

Figure 8.3 Initiation of hydrogen discharge by two types of rf coupling.

yield a more intense beam[7]; this is the mode employed in the ion source
shown in figure 8.2. For inductive coupling the absorption of energy by
the plasma is facilitated by high rf frequencies (e.g., about 100 MHz).
All other parameters remaining fixed, a given beam intensity can be obtained at lower oscillator power levels if the frequency is raised. In the
source described here, a 90 MHz oscillator was used at power levels of
up to 50 W.

A high rf frequency proved advantageous for a second reason, as well.
Underlying the theory of the atomic density matrix developed in previous
chapters is the assumption of continuous monoenergetic atom production with no preferential initial phase encountered by the beam entering
the rf spectroscopy chamber. Such conditions would be violated if the
ion source were pulsed. In a number of the early experimental investigations of inductively coupled ion sources, the extracted current was
observed to be strongly modulated (up to 100% depending on power) at
the discharge excitation frequency. However, this effect was shown to be
negligible, and the beam energy spread much less, at high rf frequencies
(~ 100 MHz) compared to low frequencies (~ 20 MHz).[8]

Protons are drawn from the source by holding the extraction electrode at a positive potential $V_{ex} \sim 1$–5 kV relative to the extraction tip.
In keeping with Wood's experiments, the anode and cathode are located
at opposite ends of the containment vessel far from the main body of the

discharge. When the system is working well, a cathode dark space forms around the extraction tip as V_{ex} is raised from zero. With continued increase of V_{ex}, the circular space contracts and eventually disappears into the exit canal, at which point the output current should be maximum.

The output current of the rf ion source depends on a number of experimental variables—source pressure, rf power, rf frequency, extraction potential, and diameter of the exit canal—all of which contribute to the shape of the meniscus plasma boundary surrounding the exit canal. This boundary is effectively the surface from which protons are emitted, and its shape is perhaps the principal single determinant of current yield. In this regard, an inconspicuous but seminal element of the ion source design is the small fused silica sleeve that isolates the cathode from the plasma. This sleeve serves two functions: it prevents electrical breakdown across the Pyrex sheath enclosing the cathode tip, and, since it is approximately at the plasma potential, it acts as a virtual anode to shape the plasma surface. Protons emitted from the boundary are accelerated across the plasma-cathode gap and focused by the electric field lines which curve into the exit canal. The critical importance of extraction geometry may be seen in the fact that small deviations in the shape of a cathode tip from the flat surface perpendicular to the canal (as shown in the figure) can drop the beam intensity by over 90%.

The potential of the ion source is raised above ground level by the high accelerating potential V_{ac} of several tens of kilovolts applied at the brass sealing flange surrounding the exit canal. To protect users from the high-voltage terminals, as well as to minimize radiofrequency radiation, the entire ion source was enclosed in a copper box with electrical connections to the oscillator (and a cooling fan) made with feedthrough capacitors. Under the conditions in which it was employed, this home-built source provided a current output of approximately 500 μA with an operation time of some 300 hours before a film of aluminum accumulated on the glass walls, and the source required cleaning. Commercially available ion sources, which employ higher rf power and a solenoid magnet (to concentrate the plasma, trap electrons, and quench plasma oscillations), can produce proton currents of several milliamperes and operate continuously for over 1000 hours.

A summary of the characteristics of the rf ion source employed by the author over a period of five years is given in table 8.1.

TABLE 8.1
Ion Source Characteristics

Nature of discharge	annular (inductive)
Gas	H_2
Output current	up to 500 μA (typical: 15–20 μA)
Percentage protons	90% (estimated)
RF power level	up to 50 W (typical: 2–10 W)
RF frequency	90 MHz
Gas consumption	1–2 cm^3/hr
Source pressure	5–10 mTorr (0.7–1.4 Pa)
Exit canal	length: 12.5 mm
	diameter: 1.0 mm
Discharge tube (Pyrex)	length: 137.5 mm
	OD: 30.0 mm
Extraction potential	1–5 kV
Acceleration potential	18–20 kV

8.3 ION ACCELERATION AND FOCUSING

Protons extracted from the ion source pass through the exit canal within a narrow cone of apex half-angle θ given approximately by the ratio of canal radius to canal length. For the geometry of figure 8.2, θ is approximately 0.05 radians, and the solid angle of emergence is $\Delta\Omega \sim \pi\theta^2 = 0.008$ steradians. Although this divergence is small, subsequent collimation is mandatory in order that the bulk of the beam pass through the apertures of the target chamber and, as nearly as possible, through the exact center of the radiofrequency spectroscopy region.

One simple and effective way to achieve this is by use of an electrostatic unipotential lens ("Einzellens" in German) constructed from three coaxial aluminum cylinders as shown in detail in figure 8.4. The distinguishing characteristic of this lens is its electrical symmetry. The two outer cylinders are kept at the same potential (ground potential in the present case), and a high potential comparable in magnitude to the accelerating potential is applied to the central element. Thus, the refractive indices of the object and image spaces are equal.[9] Geometrically, however, the lens shown in the figure is not perfectly left-right symmetric. The tapered element close to the extraction tip serves as a gap lens to

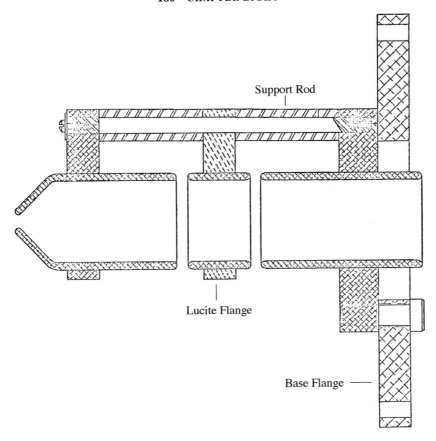

Support Rod

Lucite Flange

Base Flange ———

Figure 8.4 Details of three-element electrostatic unipotential lens.

collimate the ion beam partially by increasing the axial component of the proton velocity by a factor $\sqrt{V_{ac}/V_{ex}}$.[10]

To determine theoretically the optimum design parameters of an ion lens is effectively a two-stage problem requiring first the spatial distribution of the electrostatic potential and second the corresponding trajectory of the focused particle.

For a given lens geometry (of cylindrical symmetry), the potential distribution $V(r, z)$ is obtained by solving Laplace's equation, a procedure that in general must be implemented numerically. One such procedure, which can be performed rapidly on a computer, is the relaxation method based on the special property of harmonic functions that the value of the function at a point is equal to its average over the neighborhood of the point. The potential function is defined over an array of discrete points, and the values of interior (that is, nonboundary) points are adjusted until self-consistency is attained to within a desired margin of error. Mathematically, this is equivalent to solving the Laplace finite-difference equations (in cylindrical coordinates)

$$V_{r,z} = \tfrac{1}{4}\left(V_{r,z+1} + V_{r,z-1} + \frac{2r+a}{2r}V_{r+1,z} + \frac{2r-a}{2r}V_{r-1,z}\right)$$
$$(r \neq 0), \quad (1a)$$

$$V_{0,z} = \tfrac{1}{4}(V_{0,z+1} + V_{0,z-1} + 2V_{1,z}) \qquad (r = 0) \qquad (1b)$$

for a planar cross section divided into a mesh of small squares of length a; z is the distance along the lens axis, and r is the radial distance from the axis. The "relaxation" is the change in potential values from one iteration to the next. If the initial potential at a point is far from the exact value, this "bump" will diffuse outward through a widening neighborhood of points until smoothed to an extent permitted by the boundary conditions.[11]

Once the potential distribution is known and the initial conditions r_0 and $r_0' \equiv (dr/dz)|_{r_0}$ of the charged particle specified, the trajectory of the particle is found by solving (again numerically) the paraxial ray equation

$$r'' + \frac{V_{0,z}'}{2(V_{0,z} - V_{ac})}r' + \frac{V_{0,z}''}{4(V_{0,z} - V_{ac})}r = 0. \qquad (2)$$

Primes on the variables $r(z)$ and $V_{0,z}(z)$ indicate differentiation with respect to z. Note that the equation is independent of the charge/mass ratio of the ion, a feature characteristic of all electrostatic lenses. Moreover, to the extent that the true ion trajectory does not deviate too far from the lens axis, only the potential distribution along the lens axis is required to determine the entire trajectory in the paraxial approximation. The derivation of Eq. (2) is given in appendix 8A.

Figure 8.5 shows the potential distribution calculated by the relaxation method for the case in which the central cylinder is maintained at a positive focusing potential V_F above the two grounded outer elements. In the upper frame one sees the equipotential lines in a midsection through the lens; the complete three-dimensional distribution is obtained by rotating the figure about the lens axis. The potential is maximum at the

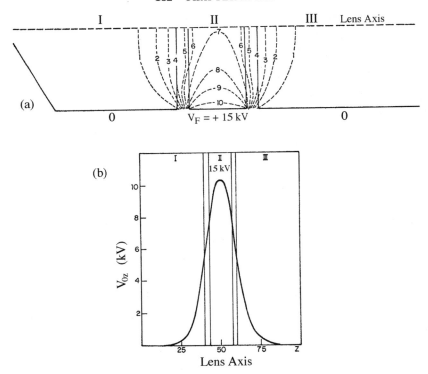

Figure 8.5 Potential distribution within a unipotential gap lens for focusing potential $V_F = +15$ kV: (a) equipotential lines in a midsection through the lens axis; numerical values represent $10\,V/V_F$; (b) potential along the lens axis. Horizontal scale: 100 units $= 16.63$ cm.

central element, falls off with decreasing r, and is symmetric about the plane bisecting the central electrode normal to the lens axis. Note that this map would look the same if the sign of the potential V_F were reversed, although the numerical values associated with the equipotential lines would be different. The lower frame shows a profile of the exact axial potential; it agrees reasonably well with the phenomenological expression $V_{0,z} = V_0 + Ae^{-Bz^2}$ which has been widely used in analytical studies of unipotential lenses.[12]

A unipotential lens can be operated in either of two modes: (a) the "decelerate-accelerate" mode in which the potential of the central element is positive with respect to that of the outer elements, and (b) the "accelerate-decelerate" mode, in which the relative potential of the central element is negative. The stated signs apply to positively charged ions;

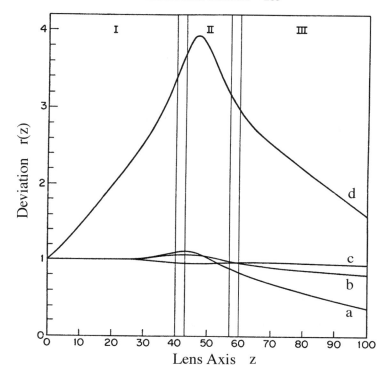

Figure 8.6 Particle trajectories through the lens for different focusing conditions: (a) $V_F = 21$ kV, $r'(0) = 0.00$; (b) $V_F = 15$ kV, $r'(0) = 0.00$; (c) $V_F = -15$ kV, $r'(0) = 0.00$; (d) $V_F = 25$ kV, $r'(0) = 0.05$. For both ordinate and abscissa 100 units = 16.63 cm.

were the lens to be used with electrons as in an electron microscope, the two modes of operation would be accomplished with potentials of the reversed sign. Figure 8.6 shows examples of proton trajectories for various values of V_F and $r'(0)$. For the same initial slope $r'(0)$, trajectories b and c illustrate that both modes of operation lead to an overall convergent beam (although the decelerate-accelerate mode provides stronger focusing). The equivalent focusing systems for light, illustrated in figure 8.7, is a double convex lens between two plano-concave lenses ($V_F = +15$ kV) for the decelerate-accelerate mode, and a double concave lens between two plano-convex lenses ($V_F = -15$ kV) for the accelerate-decelerate mode.

Experimentally, the focusing ability of the lens can be gauged from current measurements with a Faraday cup at the end of the beam line.

(a)

(b)

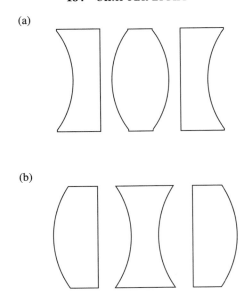

Figure 8.7 Equivalent light optical system for the (a) decelerate-accelerate and (b) accelerate-decelerate focusing modes of a symmetrical unipotential lens.

In the apparatus of figure 8.1, the Faraday cup consisted of a central circular electrode of 2 mm diameter surrounded concentrically by two annular electrodes. When focused, 80%–90% of the beam was contained in an area of about 1 mm² within the central electrode.

8.4 EXCITED ATOM PRODUCTION

In the initial rf spectroscopy experiments performed with the apparatus of figure 8.1, hydrogen atoms were created and excited by electron-capture collisions of the proton beam with a carbon foil target a few hundred atoms thick stretched across a metal frame at the center of the collision chamber. Earlier research had established that a substantial fraction of the incident protons captured an electron to emerge from the foil as fast-moving neutral atoms. Since the interaction time was extremely short—a 20 keV proton spending only about 10^{-14}–10^{-13} seconds in passage through a foil of thickness 10 μg/cm²—the uncertainty principle ($\Delta E \, \Delta t \geq \hbar$) predicts that emerging atoms will be distributed widely over their energy eigenstates. Indeed, under appropriate circumstances foil-excited atoms were produced, not in an incoherent mixture, but in a coherent linear superposition of states, as evidenced by the appearance of "quantum beats" (modulated exponential decay) in the spontaneous emission.[13]

Although possessing the advantage of a well-defined point of interaction, carbon foils were scarcely able to sustain a beam current greater than 5 μA, and their progressive deterioration led to a decreasing signal with final rupture often occurring before the end of an experimental run. The foil targets were eventually replaced by a more robust gas target which permitted larger beam currents and led to higher signal stability.

An enlarged view of the gas target chamber, in which such gases as H_2, N_2, He, Ar, Xe, and others served as targets, is shown in figure 8.8. Four steel support rods, which screw into the top (fore and aft) and sides (front and rear—not shown) of the chamber, and pass through the walls of the target housing by means of O-ring seals, permit fine adjustment of the target orientation. At a target pressure of about 5 mTorr single-collision conditions were obtained with an energy dispersion of less than 100 eV out of 20 keV.[14] This is to be compared with an energy loss of 2–10% for proton passage through a carbon foil target.

The length of the interaction region of a gas target is dictated by the lifetimes of the excited atomic states to be investigated. In the design shown in figure 8.8, an interaction region of 10 cm was selected to allow study of the moderately excited states of the hydrogen atom. For example, moving at a speed of 2×10^8 cm/s, a 20 keV proton crosses the gas target in 50 ns, a time sufficiently short to permit probing of all but the shortest lived H-atom P-states, as indicated in table 8.2.[15]

The lifetime of states of given orbital angular momentum lengthens with increasing principal quantum number n. On the other hand, the electron-capture cross section leading to production of these states diminishes roughly with n^3 and is largely independent of the nature of the gas for targets with atomic number $Z > 7$. A noteworthy exception to the Z-independence of the capture cross section is the production of $H(2s)$ states by bombardment of Cs vapor with protons of energy less than 10 keV. As a result of an accidental resonance, the resulting cross section exceeds those of other target materials by two orders of mag-

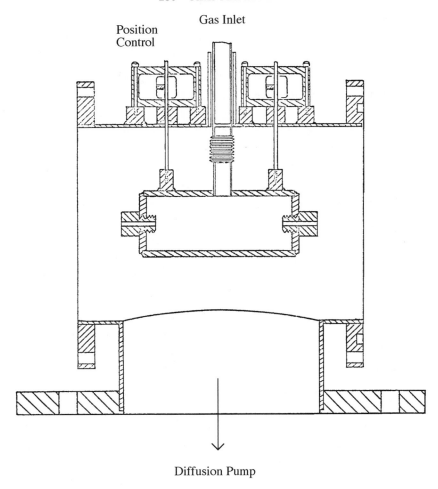

Figure 8.8 Cross section through the gas target.

nitude. A Cs vapor target, however, is not as easily constructed as the much simpler one for common diatomic or noble gases.

When in operation, the flow of target gas was controlled by a variable leak valve so as to maintain a pressure of approximately 5 mTorr inside the target chamber, as monitored by a Pirani gauge. Under these conditions the degree of neutralization of the proton beam was about 50%.

TABLE 8.2
Hydrogen Lifetimes

State	Lifetime (ns)
2s	0.14 seconds
2p	1.6
3s	160
3p	5.4
3d	15.6
4s	230
4p	12.4
4d	36.5
4f	73
5s	360
5p	24
5d	70
5f	140
5g	235

8.5 THE RADIOFREQUENCY SYSTEM

In the apparatus illustrated in figure 8.1, each rf field with which the atomic beam interacts is produced by a parallel-plate capacitor inserted into a 50 Ω coaxial transmission line. Several such capacitors were constructed to allow different interaction times, to cover different frequency ranges, or to serve as either a switched spectroscopy field or a non-switched state-selection field. Recall from earlier chapters that atomic transitions are monitored by the difference in photon counts when the rf spectroscopy field is "on" compared to when it is "off."

Figure 8.9 gives a block diagram of the rf system as it was first employed. The rf power, generated over a range of frequencies (from about 60 MHz to 4 GHz in the first experiments on hydrogen) by a family of oscillators, passes through a directional coupler that channels part of the power to a frequency counter and the rest to the input terminal of a solid-state (single-pole single-throw) switch. The switch is activated by pulses from a gating signal generator that synchronously controls the rf-on and rf-off counters. When the switch bias is positive, no rf passes and the rf-off counter is gated on. With a reverse bias, the rf-on counter

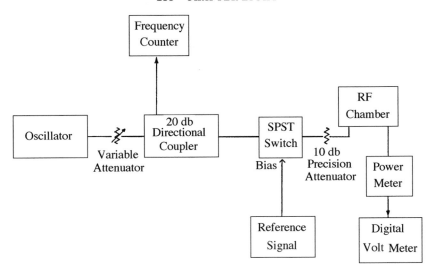

Figure 8.9 Block diagram of the radiofrequency system.

is activated, and the power is passed through the rf chamber to a 50 Ω termination provided by the thermistor mount of a power meter.

Under ideal circumstances the rf power is transmitted unreflected through the capacitor as a single-mode transverse electromagnetic (TEM) wave with an instantaneous field configuration identical to that of a static electromagnetic field with the same boundary conditions. (For most experiments employing these wave guides, the rf wavelength far exceeded the capacitor length.) In an ideal parallel-plate capacitor the static electric field is uniform across the interaction region. The rf field will therefore display the simple harmonic time dependence and spatial homogeneity that was assumed throughout the theoretical analysis in earlier chapters. Model conditions do not always hold, however, in the design and use of real components. In the present case, for example, nonideal performance can occur as a result of the finite extent of the capacitor and the impedance mismatch between the connectors and coaxial transmission line.

Consider first the matter of finite capacitor length. As is well known, the field configuration of a real parallel-plate capacitor is not perfectly uniform but extends beyond the plates as a "fringing" field. This fringing field is inhomogeneous, contains both transverse components (perpendicular to the plates) and longitudinal components (parallel to the plates), and diminishes rapidly with distance from the capacitor. The extension of the rf field beyond the plates results in an uncertainty in the atom-field interaction time. Furthermore, the longitudinal components of the field can couple Zeeman states (defined with respect to the transverse field as the axis of quantization) of different M_F quantum number. For transitions between the components of two hyperfine levels [e.g., $n^2S_{1/2}(F = 1, M_F) \leftrightarrow n^2P_{1/2}(F = 1, M'_F)$] in the absence of a magnetic field, the only effect is a line broadening, since the dipole matrix elements between different Zeeman substates may be different. There is no frequency shift, however, because of substate degeneracy within each hyperfine level. If, as is more often the case, transitions between different pairs of hyperfine levels overlap, an asymmetric shift can result.

Another possible consequence of the fringing field is an analog of the so-called Millman effect, a resonance asymmetry that arises if the direction of the oscillatory field varies along the length of the interaction region.[16] Also referred to as the "hairpin" effect, the asymmetry was first noted in early magnetic resonance experiments in which the source of the oscillating magnetic field was a hairpin-shaped wire loop. In the present case, the instantaneous direction of the fringing field intersecting a plane parallel to the capacitor plates rotates continuously as a function of distance from the capacitor. Thus, an atom with speed v crossing a point at which the direction of the field changes with distance at a rate $d\theta/dz$ will experience two rf fields whose frequencies are $\Delta\omega = v(d\theta/dz)$ above and below the basic oscillator frequency ω. This can be understood by resolving the linearly polarized rf field into two oppositely rotating components. One of these components rotates in the sense of the incremental frequency $\Delta\omega$; the other component rotates in the opposite sense.

To ascertain whether the preceding effects are significant or not, it is necessary to know the exact electric field configuration experienced by the atoms in the beam. This was done for the rf system described here by employing again a finite-difference relaxation method whereby Laplace's equation was solved over a 100×100 element grid with one capacitor plate at a specified positive potential and the other plate, together with the grid boundary, at ground potential. The method, analogous to that used in the design of the unipotential lens, yielded the spatial distribution

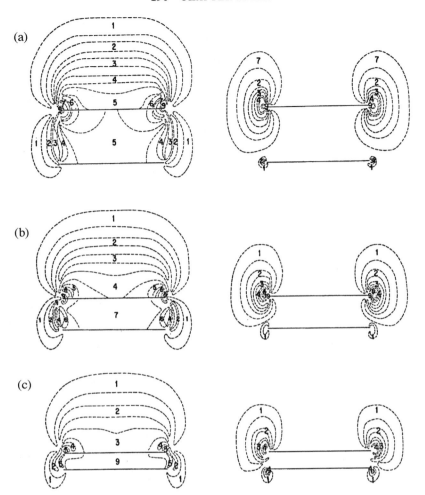

Figure 8.10 RF field contour maps for capacitors of different ratios of plate length to separation: (a) 3:1, (b) 5:1, (c) 10:1. Transverse-field maps are on the left; longitudinal-field maps are on the right. For a potential difference of 2 V across plates separated by 1 cm, the numerical value of the field in each coded region is obtained by multiplying the code number by (a) 0.404 (transverse field), 0.545 (longitudinal field); (b) 0.299, 0.353; (c) 0.222, 0.208.

of the potential, as well as the transverse and longitudinal components of the electric field.

Figure 8.10 shows rf field contour maps for different ratios of plate length to plate separation. As expected, the larger the ratio, the closer a capacitor approaches the homogeneity of the ideal parallel-plate capaci-

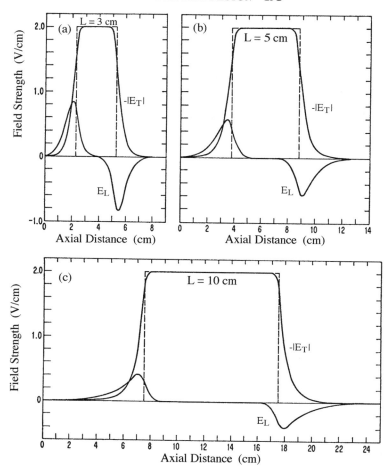

Figure 8.11 Transverse (E_T) and longitudinal (E_L) fields (solid lines) along the beam axis for the same capacitors as in figure 8.10. Dashed lines show the step potential of an ideal capacitor.

tor. For all the capacitors modeled, however, the transverse electric field was reasonably uniform in the midplane containing the beam trajectory. Figure 8.11 shows explicitly the variation in transverse and longitudinal fields encountered by an atom moving along the symmetry axis of the midplane.

Once the exact electric field configuration is known, the resulting resonance lineshapes can be calculated for desired atomic transitions by nu-

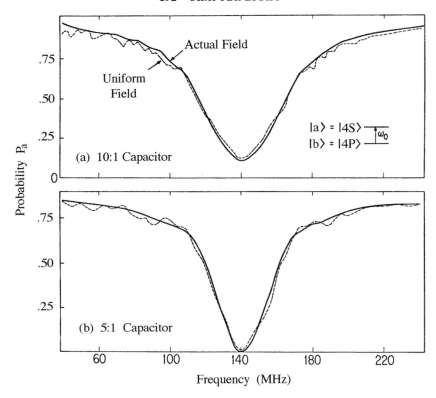

Figure 8.12 Two-level resonance lineshapes computed numerically for a uniform field and for the actual transverse fields within 10:1 and 5:1 capacitors. Theoretical parameters are decay rates $\gamma_a = 4.35 \times 10^6$ s^{-1}, $\gamma_b = 80.7 \times 10^6$ s^{-1}; resonance frequency $\omega_0/2\pi = 140$ MHz; interaction strength (of uniform field) $V_{ab} = 9.92$ MHz; proton energy 20 keV.

merically integrating the Schrödinger equation over the beam trajectory. Such analyses for selected hydrogen transitions have established that the actual field configuration within capacitors of length-to-separation ratios as small as 5:1 yields resonance profiles in excellent agreement with those computed numerically for an ideally uniform transverse field. As shown in chapter 3, these exact lineshapes are reliably reproduced by the generalized resonant field (GRF) theory. Contributions from the longitudinal component of the fringing fields are for the most part negligible.

An example of such a lineshape comparison is reproduced in figure 8.12 for the case of a $4^2S_{1/2}$–$4^2P_{1/2}$ transition. Although small de-

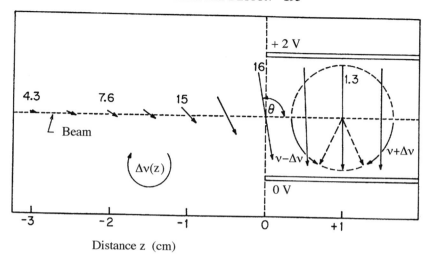

Distance z (cm)

Figure 8.13 Electric Millman effect. The magnitude and direction of the fringing field produced by a 10 cm capacitor is shown as a function of the distance along the beam. Numbers above field lines indicate the instantaneous incremental frequency (in MHz) for a 20 keV H atom. At a point 1 cm inside the capacitor, the net field is shown resolved into two oppositely rotating components.

viations between predictions based on a uniform field and those of the actual field occur in the wings of the profile, the two lineshapes overlap nearly identically in the region of resonance; in particular, there is no shift in the resonance location, a point of importance for the determination of atomic level separations such as that of the Lamb shift.

From the results of the relaxation studies, an accurate appraisal of the electric Millman effect can also be made. Figure 8.13 shows the instantaneous magnitude and orientation of the rf field produced by a 10 cm capacitor along the trajectory of a 20 keV atom (with speed $v \sim 2 \times 10^8$ cm/s). The incremental frequency in MHz is indicated periodically along the trajectory, and the field rotation is seen to be clockwise (regardless of the relative plate potential). Since the transition probability is proportional to the square of the electric field strength, the Millman effect is most significant within about the first centimeter of the rf plates; at any point within this region the atom experiences fields of angular frequencies $\omega \pm \Delta\omega$ where the incremental frequency $\Delta\omega$ decreases rapidly with penetration of the capacitor. The result is that the beam senses a rf field with a slightly broader frequency spectrum, an effect contributing to the width of a resonance line, but not to a line shift or to the creation of satellite lines. Unlike the magnetic Millman effect, both circular polarizations contribute to the lineshape because degenerate magnetic substates of both $\pm M_F$ quantum numbers are present.

In general, it is probably safe to say that for spectroscopy with a *single* rf field, the effects of finite capacitor size on lineshape are unimportant beyond a certain plate length-to-separation ratio. For separated coherently oscillating fields, however, where fringing from two rf capacitors can penetrate the "field-free" region, the problem is likely to be more serious, especially for short capacitors.

Consider next the effects of impedance mismatch between the transmission line (from the oscillator) and the rf chamber, leading to standing waves in the interaction region. As an illustrative first approximation, which will be developed more thoroughly in appendix 8B, the total field within the chamber can be written as

$$
\begin{aligned}
E(t, z) &= E^{(-)}(t, z) + E^{(+)}(t, z) \\
&= \left[E^{(-)}e^{-i\omega(t-z/c)} + \text{c.c.}\right] + \left[E^{(+)}e^{-i\omega(t+z/c)} + \text{c.c.}\right]
\end{aligned} \quad (3)
$$

with $E^{(-)}(t, z)$ the incident wave generated by the laboratory oscillator and $E^{(+)}(t, z)$ the reflected wave. The rf chambers are designed so that the incident wave propagates either parallel to the beam (as shown in figure 8.1) or perpendicular to the beam. Longer interaction times can be obtained with the first configuration, whereas in the second configuration the Doppler shift is of second order. For the present discussion, we assume that the incident wave travels with the beam. Rewriting Eq. (3) in the form

$$
E(t, z) = e^{-i\omega(t-z/c)}E^{(-)}\left(1 + \rho e^{-2i\omega z/c}\right) + \text{c.c.} \quad (4)
$$

with complex reflection coefficient

$$
\rho = |\rho|e^{i\phi} = \frac{E^{(+)}}{E^{(-)}}, \quad (5)
$$

and substituting $z = vt$, we obtain

$$E(t) = E^{(-)}e^{-i\omega^{(-)}t} + \rho E^{(-)}e^{-i\omega^{(+)}t} + \text{c.c.} \qquad (6a)$$

with

$$\omega^{(\pm)} = \omega \pm \beta\omega, \qquad \beta = v/c, \qquad (6b)$$

for the total field experienced by the atom at time t.

The atom is therefore immersed in two electric fields of different frequency and strength, with the consequence that, depending on the reflection coefficient, resonance lineshapes may suffer shifts and distortions not predicted by the theoretical analysis of the earlier chapters. In appendix 8B the resolvent operator formalism is used to determine the lineshape under the conditions of impedance mismatch. For a weak reflected wave ($|\rho| < 1$), it is shown that the primary effect is a small asymmetric shift.

A second consequence of impedance mismatch is that, due to the reflected wave, the power level registered by the thermistor will be less than that within the interaction region. If this discrepancy is frequency dependent, the recorded lineshapes may again be spuriously shaped. The electric field inside the rf chamber is not an experimentally accessible quantity; only the power transmitted through the chamber and received at the terminal load (the thermistor) is actually measurable. To handle this problem, it is useful to model the rf chamber phenomenologically by considering each connector as a two-port device inserted into the transmission line. The properties of the rf chamber can then be characterized by elements of a scattering matrix—one matrix relation for each connector—and, by exploiting geometrical and material symmetries, a correlation can be established between the power experienced by the atomic beam and that delivered to the load.

The details of this model, leading to an expression for the actual power in the chamber in terms of the reflection coefficient, are summarized in appendix 8D. For rf chambers discussed in this chapter, the magnitude of the reflection coefficient was found to be very small—roughly 0.01 to 0.05—over the range of applied frequencies. Thus power deviation in the experiments performed with the apparatus was insignificant.

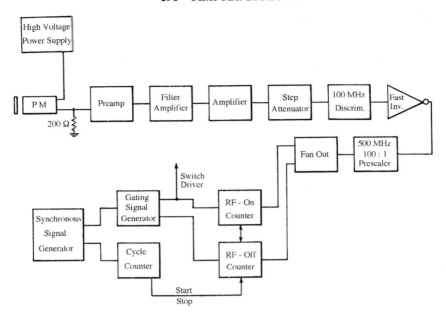

Figure 8.14 Optical detection and signal processing system. Wavelength-selected photons are counted for equal "on" and "off" periods of the rf spectroscopy field. The signal is proportional to the difference in counts.

8.6 OPTICAL DETECTION

A schematic diagram of the optical detection and signal processing systems is shown in figure 8.14. Atoms, after interacting with one or more rf fields, exit the last rf chamber and radiatively decay in passage through the optical detection chamber (on their way to the beam dump). Single photons, wavelength-selected by a narrow bandpass interference filter, enter an end-on photomultiplier tube with appropriate photosensitive cathode. In the spectroscopy of atomic hydrogen, for example, Balmer radiation was detected by a tube containing a trialkali cathode with quantum efficiency of 18–21% in the blue-violet (H_β, H_γ) and 6% in the red (H_α). A semicylindrical mirror inserted at the rear of the chamber enhances the amount of light (of all polarizations) received at the detector.

After passing through several wide-band amplifiers, the optical signal is ultimately channeled equally to the rf-on and rf-off counters. These scalers operate synchronously with the solid state switch controlling passage of rf power into the spectroscopy chamber. In the apparatus of figure 8.1, the counting sequence was initiated and terminated by a presentable cycle counter which received instructions from a signal generator that synthesized time intervals of 1 to 1000 ms in 1 ms steps from a 100 MHz standard frequency. The gating times of each counter were equal to within one part in ten million.

8.7 SPECTROSCOPY

In the applications to atomic hydrogen spectroscopy illustrated in this section, the fast hydrogen atoms were generated from an initial 10–20 μA proton beam in the energy range of 15–20 keV, with 18 keV the setting for most of the experiments. With the rf interaction time fixed by beam velocity and rf capacitor length, the electric field strength leading to an optimum signal for a given transition can be determined theoretically by solution of a transcendental equation like those derived in chapter 2. [See Eqs. (2.47a) and (2.47b).] In practice, however, a compromise between large signals and narrow linewidths was achieved by setting the operating power level empirically at the half-saturation point determined from a plot of signal intensity versus power (at a specified resonance frequency), like that of figure 8.15. Once established, the power level is maintained constant throughout the recording of the rf lineshapes in a given run.

Figure 8.16 presents examples of panoramic sweeps of fine structure transitions in the $n = 3, 4, 5$ electronic manifolds of hydrogen. No quenching or state-selecting fields were employed, and each fine structure lineshape is a complex of three hyperfine components according to the selection rule $\Delta F = 0, \pm 1$, transitions between two $F = 0$ states being disallowed. The signal, as defined by Eq. (6.42), is proportional to the difference in photon counts (of Balmer radiation) in the absence and presence of the rf field, respectively.

The H($n = 3$) resonances shown in figure 8.16a were recorded with a 3 cm rf capacitor at a power level of 20 mW; excited atoms were generated by proton collisions with a N_2 target. Except for the weak $3^2D_{3/2}$–$3^2P_{1/2}$ transitions, the fine structure resonances are quite clearly resolved. Figure 8.16b shows resonances of H($n = 4$) atoms generated

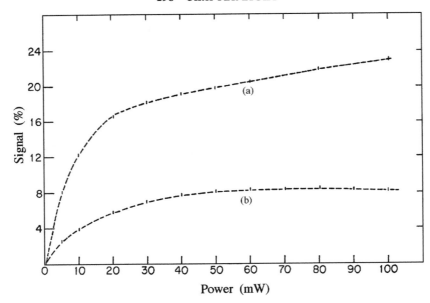

Figure 8.15 Power saturation curves for the $4^2S_{1/2}-4^2P_{1/2}$ transition under experimental conditions: (a) frequency 130 MHz, field length 9 cm, N_2 target; (b) frequency 140 MHz, field length 3 cm, Ar target.

by an argon target interacting with a 10 mW field of length 9 cm. F states make their first appearance in the $n = 4$ manifold. Since these states do not contribute to Balmer radiation, the negative region of the $F–D$ resonances signifies that the rf field has increased the net D state population. Fine structure profiles are fairly well resolved except for overlapping $4^2F_{5/2}-4^2D_{3/2}$ and $4^2D_{5/2}-4^2P_{3/2}$ transitions and the submergence of the $4^2D_{3/2}-4^2P_{1/2}$ resonance in the high-frequency wing of the $4^2P_{3/2}-4^2S_{1/2}$ lineshape. Figure 8.16c exhibits H($n = 5$) spectra in which excited atoms, produced by a N_2 target, passed through a 5 cm interaction region of 15 mW. Here there is considerable overlap of resonances. Although the $n = 5$ manifold marks the first apperance of G states, no $F–G$ transitions can be observed directly in the Balmer γ radiation since neither state can couple to the $n = 2$ manifold via a single-quantum electric dipole transition.

In figure 8.17 we see in more detail the composition of a fine structure transition—in this case $3^2S_{1/2}-3^2P_{1/2}$. As shown in the insert, the total lineshape results from three allowed hyperfine transitions $(S_{1/2})F \leftrightarrow$

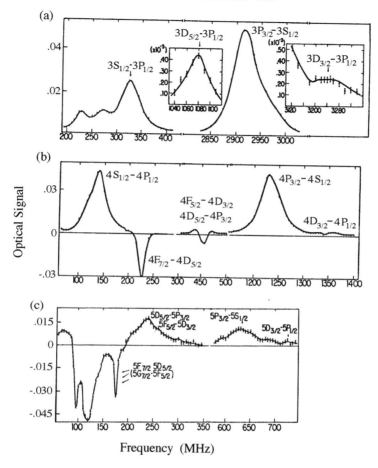

Figure 8.16 Panoramic sweeps of H($n = 3, 4, 5$) transitions. (a) H($n = 3$), N_2 target, field length $L = 9$ cm, field strength $E_0 = 1.41$ V/cm, direction $\theta = \pi$; (b) H($n = 4$), Ar target, $L = 9$ cm, $E_0 = 1.00$ V/cm, $\theta = \pi$; (c) H($n = 5$), N_2 target, $L = 9$ cm, $E_0 = 1.23$ V/cm, $\theta = \pi$.

($P_{1/2}$)F': (a) $0 \leftrightarrow 1$, (b) $1 \leftrightarrow 1$, (c) $1 \leftrightarrow 0$, in order of increasing frequency. Although the exact lineshapes for these transitions can be calculated precisely from the two-state theory of chapter 2 (the initial P-state amplitude being essentially zero in the rf interaction region), it is at times convenient simply to fit the lineshape to a sum of Lorentzian profiles in order to estimate the Bohr frequency and relative weighting of the three hyperfine contributions. The full curve was obtained from a least-square procedure in which the calculated parameters include the fractional con-

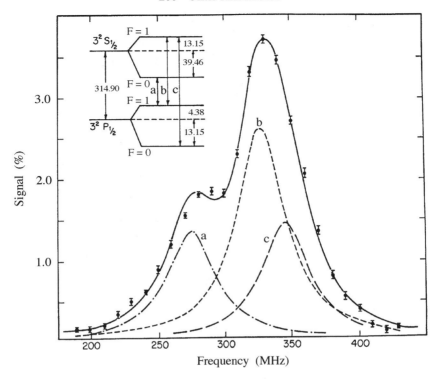

Figure 8.17 Detailed composition of $3^2S_{1/2}$–$3^2P_{1/2}$ lineshapes obtained by a least-squares fit to three Lorentzian profiles. Hyperfine transitions are identified in the inserted level diagram.

tribution of each component (dashed curves), linewidth (same for each profile), and location of the 1 ↔ 1 component; the locations of the other two lines were determined from the theoretical hyperfine separations. This figure illustrates one of the initial runs in which a carbon foil target was used to create the excited hydrogen atoms. It is worth mentioning here that the contributions of the three hyperfine components are in the ratio 1:2:1, as expected on the basis of the electron-capture model to be discussed in the next section.

As an illustration of the use of sequential rf fields to resolve spectra, figures 8.18 and 8.19 show examples of fine structure and hyperfine structure state selection within the H($n = 4$) manifold. The first figure is a panoramic sweep of the H($n = 4$) manifold under experimental conditions different from those of figure 8.16. The amplitude of the

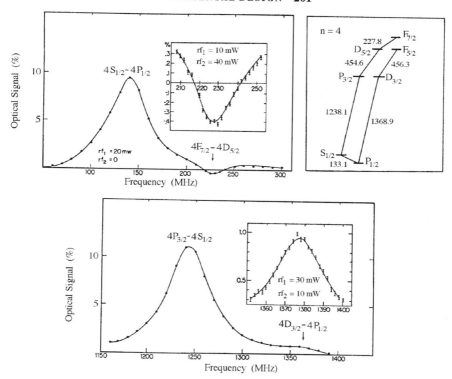

Figure 8.18 Panoramic sweep of H($n = 4$) resonances showing examples of fine structure state selection. Use of quenching fields (40 mW at 140 MHz; 30 mW at 1225 MHz) to remove $4^2S_{1/2}$ states enhances the resolution of $4^2F_{7/2}$–$4^2D_{5/2}$ and $4^2D_{3/2}$–$4^2P_{1/2}$ lineshapes, respectively.

$4^2F_{7/2}$–$4^2D_{5/2}$ lineshape is now weaker, and the profile arising from overlapping $4^2F_{5/2}$–$4^2D_{3/2}$ and $4^2D_{5/2}$–$4^2P_{3/2}$ transitions is now positive and no longer oscillatory. Moreover, the $4^2D_{3/2}$–$4^2P_{1/2}$ transitions show up less visibly than before as merely an elongation of the right wing $4^2P_{3/2}$–$4^2S_{1/2}$ of the lineshape. However, upon application of a 30 mW quenching field set at 1225 MHz to remove $4^2S_{1/2}$ states, the $4^2D_{3/2}$–$4^2P_{1/2}$ lineshape is greatly enhanced. Similarly, quenching of $4^2S_{1/2}$ states by a 40 mW field set at 140 MHz heightens the visibility of the $4^2F_{7/2}$–$4^2D_{5/2}$ lineshape.

The use of hyperfine state selection is shown in figure 8.19 for the case of the $4^2S_{1/2}$–$4^2P_{1/2}$ resonance. Here a 20 mW quenching field set at 115 MHz has removed nearly all $4^2S_{1/2}(F = 0)$ states by driving the

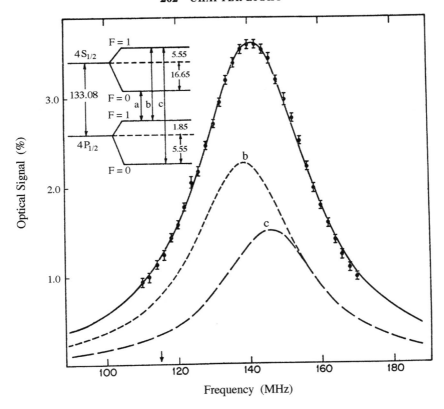

Figure 8.19 Example of hyperfine state selection. Application of a 20 mW rf field at 115 MHz has quenched nearly all $4^2S_{1/2}(F = 0, M_F = 0)$ states, thereby eliminating transition a $(0 \leftrightarrow 1)$, as confirmed by a least-squares fit of three Lorentzian curves to the composite $4^2S_{1/2}$–$4^2P_{1/2}$ resonance.

$0 \leftrightarrow 1$ transition before the beam enters the spectroscopy chamber. A least-squares fit of three Lorentzians (dashed lines) to the observed lineshape (full line) shows that only transitions originating from $4^2S_{1/2}(F = 1)$ states contribute.

When the highest degree of precision is required—as in spectroscopic tests of quantum electrodynamics—separated coherently oscillating fields can be employed to narrow a lineshape beyond that achievable with a single rf field. Figure 8.20 illustrates this narrowing in the case of the $3^2S_{1/2}$–$3^2P_{1/2}$ resonance.[17] After rf quenching to remove $3^2S_{1/2}(F = 1)$ states from the beam, the remaining $3^2S_{1/2}(F = 0)$ atoms traversed a rf interaction region in 15.5 ns, a field-free region in 18 ns, and finally

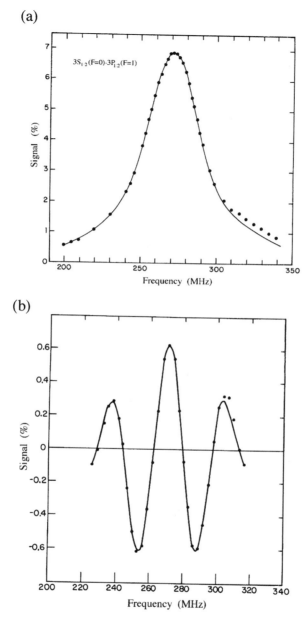

Figure 8.20 Separated oscillatory field measurement of the hyperfine-selected $3^2S_{1/2}(F = 0)–3^2P_{1/2}(F = 1)$ transition. Experimental conditions: interaction time = 15.5 ns; field-free transit time = 18 ns; power = 40 mW. (a) Full line-shape for relative phase $\Delta = 0$. (b) Isolation of quantum interference effect in the signal difference $S_1(0, \pi)$ for phases $\Delta = 0$ and $\Delta = \pi$. Solid curves are theoretical lineshapes.

a second interaction region of the same length as the first. Part (a) of the figure shows the optical signal obtained when the two rf fields are coherently oscillating in phase ($\Delta = 0$); the lineshape resembles that of a single-field experiment.

The curve shown in part (b) is the difference in optical signals at each rf frequency for in-phase and out-of-phase ($\Delta = \pi$) oscillations of the two rf field regions. As detailed in the preceding chapter, subtraction accentuates the quantum interference arising from the indeterminate location (field 1 or field 2) at which an S–P transition occurs. Narrowing of the interference pattern also occurs as a consequence of the fact that atoms surviving passage through the field-free region have a longer-than-average lifetime and hence smaller energy uncertainty. The width of the central fringe in figure 8.20 is about one-quarter the width of the corresponding profile produced with a single rf field and about one-half the width of the interference pattern made with two coherently oscillating fields separated by the much shorter field-free transit time of 1 ns.

Although generating resonance lineshapes by varying frequency rather than a static magnetic field eliminates many of the systematic errors enumerated in the preface, there are two effects that need be mentioned to insure a meaningful comparison of theory and experiment. These are the rf Doppler shift and the electrostatic Stark effect arising from stray electric fields.

8.7.1 Doppler Effect

As discussed in chapter 1, a frequency ω rads/s generated by a laboratory oscillator will be observed in the atomic rest frame to occur at ω' given by

$$\omega' = \gamma\omega\left(1 - \frac{v}{c}\cos\theta\right),$$

$$\gamma = \frac{1}{\sqrt{1 - \left(\dfrac{v}{c}\right)^2}}, \tag{7}$$

where the rf wave propagates through the interaction chamber at an angle θ to the beam. The chambers of section 8.5 were designed so that θ can be 0, $\pi/2$, or π radians. For $\theta = 0$ or π and a beam of atoms moving at speed v low compared to that of light, there results a first-order Doppler shift of the location ω_0 of each rf resonance by $\pm(v/c)\omega_0$, respectively. Although this shift can be moderately large (e.g., 10 MHz for a 1000 MHz resonance if $v/c \sim 1\%$), the Doppler effect does not pose a limit to attainable precision, since the beam velocity is well defined. When $\theta = \pi/2$, there is a second-order shift of $-\frac{1}{2}(v/c)^2\omega_0$ which is ordinarily negligible unless the atoms are moving at relativistic speeds in the laboratory. Similarly, no correction need be made for the dilatation of atomic lifetimes, which is a second-order effect.

8.7.2 Stray Fields

In the course of collecting data over a sufficiently long period, deposition of pump oil within the rf chambers can produce a gradual diminution in signal amplitude and shifting of line positions. These anomalies, attributable to the Stark effect from stray electric fields, can be eliminated by cleaning the rf plates. To estimate an upper limit to the magnitude of stray electric fields, one can exploit the extreme sensitivity of overlapping transitions between nearly degenerate pairs of states, such as $(F_{5/2}, D_{5/2})$ and $(D_{3/2}, P_{3/2})$, the separation of which is almost entirely quantum electrodynamical in origin. From the theory of chapter 2 [see Eq. (2.26)], it follows that two nearly degenerate states are thoroughly mixed at a field strength given approximately by $|V_{12}| \approx \frac{1}{4}|\gamma_1 - \gamma_2|$. For example, a field of about 0.42 V/cm is required to mix $4^2D_{3/2}$ and $4^2P_{3/2}$ states. A field of this magnitude, however, would displace the $4^2S_{1/2}$–$4^2P_{1/2}$ resonance frequency by about 0.5 MHz, which should far exceed any observed discrepancy under normal operating conditions. One can conclude, therefore, that so long as no change is observed in the $4^2F_{5/2}$–$4^2D_{3/2}$ and $4^2D_{5/2}$–$4^2P_{3/2}$ complex, the stray field within the interaction region is considerably weaker than 0.4 V/cm.

The sensitivity of atoms to electric fields increases rapidly with the degree of excitation. For example, in hydrogenic states of high principal quantum number n (Rydberg states), the electronic polarizability increases approximately as n^6. Experiments with atoms excited to levels with $n \sim 500$ have permitted the detection of stray electric fields as small as a few tens of microvolts per cm.[18]

8.8 ELECTRON CAPTURE AND ATOM FORMATION

Charge-exchange processes of the kind $X^+ + Y \to X + Y^+$ play an important role in atomic physics, atmospheric physics, astrophysics, the physics of plasmas, and elsewhere in physics and chemistry. Although the principal objective of rf spectroscopy has long been to provide the Bohr frequencies—and therefore insight into the internal interactions— of atoms and molecules, it can also be used, as discussed in earlier chapters, to study the relative initial populations of excited states coupled by the applied oscillating field. In this regard, rf lineshapes aid in understanding the electron-capture collisions that convert a proton beam into a beam of fast neutral H atoms.

To apply the lineshape theory developed in earlier chapters to the problem of determining relative state populations—in effect, the initial density matrix of an incoherently produced beam—we need a model of the electron-capture interaction that expresses the sought-for matrix elements in terms of a manageable number of parameters. One approach is the Coulombic spin-independent collision (CSIC) model[19] in which it is assumed that single electron capture involves only orbital (rather than spin) degrees of freedom. At the initial instant ($t = 0$), a proton collides with a target atom resulting in the capture of an outer target electron. The state vector of the resulting hydrogen atom is a superposition of field-free orbital (L), electron spin ($S = \frac{1}{2}$), and nuclear spin ($I = \frac{1}{2}$) basis states (referred to as the uncoupled angular momentum or u.a.m. basis),

$$|\Psi_0\rangle = \sum_{\substack{n, L \\ M_L, M_S, M_I}} Q_n^{LM_L} |nLM_L\rangle |SM_S\rangle |IM_I\rangle, \qquad (8)$$

weighted by a complex amplitude, $Q_n^{LM_L}$, the magnitude of which depends only on principal, orbital, and orbital substate quantum numbers. In the absence of external fields, and with the quantization axis defined by the direction of the beam, the creation of atomic states should be independent of the sign of the magnetic quantum numbers; thus states of the same L, $|M_L|$ in the expansion (8) must be weighted equally:

$$|Q_n^{LM_L}| = |Q_n^{L\bar{M}_L}|. \qquad (9)$$

The density matrix for the initial state of an ensemble of similarly prepared hydrogen atoms is then obtained by substituting state vector (8) into Eq. (3.3) defining the statistical operator:

$$
\sigma^o = \langle |\Psi_0\rangle\langle\Psi_0|\rangle_\phi
$$
$$
= \sum_{\substack{n, n', L, L' \\ M_L, M_{L'}, M_S, M_{S'}, M_I, M_{I'}}} |nLM_LM_SM\rangle \sigma^{o\,nLM_LM_SM_I}_{\;\;\;n'L'M_{L'}M_{S'}M_{I'}}\langle n'L'M_{L'}M_{S'}M'|,
$$

(10)

where [from Eq. (3.4a)]

$$
\sigma^{o\,nLM_LM_SM_I}_{\;\;\;n'L'M_{L'}M_{S'}M_{I'}} = \langle Q_n^{LM_L}(Q_{n'}^{L'M_{L'}})^*\rangle_\phi
$$

(11)

is a density matrix element in the u.a.m. representation.

Under the assumption that there are no specific nonrandom phase relationships among the amplitudes, phase averaging in Eq. (11) leads to elements of the form

$$
\sigma^{o\,nLM_LM_SM_I}_{\;\;\;n'L'M_{L'}M_{S'}M_{I'}} = \sigma^{LM_L}_n\delta_{nn'}\delta_{LL'}\delta_{M_LM_{L'}}\delta_{M_SM_{S'}}\delta_{M_IM_{I'}}
$$

(12a)

with

$$
\sigma^{LM_L}_n \equiv |Q^{LM_L}_n|^2,
$$

(12b)

and therefore a statistical operator with only diagonal elements

$$
\sigma^o = \sum_{\substack{n, L \\ M_L, M_S, M_I}} |nLM_LM_SM\rangle \sigma^{LM_L}_n\langle nLM_LM_SM|
$$

(13)

in the u.a.m. basis.

The absence of coherence terms in the initial density matrix (13) does not mean, however, that there is no coherence in the physical states of the H atom. The states upon which spectroscopy is performed and which are detected via their decay radiation are eigenfunctions of the field-free Hamiltonian including spin-orbit and hyperfine interactions. We must therefore re-express the initial density matrix in the coupled angular momentum (c.a.m.) representation. This is accomplished by the projection operation

$$
\sigma^{o\,nLJFM_F}_{\;\;\;n'L'J'F'M_{F'}} = \mathrm{Tr}\{\sigma^o|nLJFM_F\rangle\langle n'L'J'F'M_F'|\}.
$$

(14)

The transformation between u.a.m. and c.a.m. bases is effected by the Clebsch-Gordan coeffciicents [see Eq. (3.51)] and leads to the nonvanishing elements

$$\sigma^{\mathrm{o}nLJFM_F}_{nLJ'F'M_F} = \sum_{M_L, M_S, M_J, M_I} \sigma^{LM_L}_n \langle LSM_LM_S \mid JM_J\rangle\langle LSM_LM_S \mid J'M_J\rangle$$
$$\times \langle JIM_JM_I \mid FM_F\rangle\langle J'IM_JM_I \mid FM_F\rangle, \tag{15}$$

where there is only one contribution to the sum over M_J, namely, $M_J = M_L + M_S = M_F - M_I$.

Equation (15), rewritten in terms of the more symmetrical 3-J symbols [Eq. (3.53)], becomes

$$\sigma^{\mathrm{o}nLJFM_F}_{nLJ'F'M_F} = \sum_{M_L, M_S, M_I} (-1)^{J+J'+2M_J} \sigma^{LM_L}_n$$
$$\times \left\{(2J+1)(2J'+1)(2F+1)(2F'+1)\right\}^{1/2}$$
$$\times \begin{pmatrix} L & S & J \\ M_L & M_S & -M_J \end{pmatrix} \begin{pmatrix} L' & S & J' \\ M_L & M_S & -M_j \end{pmatrix}$$
$$\times \begin{pmatrix} J & I & F \\ M_J & M_I & -M_F \end{pmatrix} \begin{pmatrix} J' & I & F' \\ M_J & M_I & -M_F \end{pmatrix}. \tag{16}$$

From the properties of the 3-J symbols follows a simple relationship between density matrix elements differing only in the sign of the magnetic quantum number:

$$\sigma^{\mathrm{o}nLJF\bar{M}_F}_{nLJ'F'\bar{M}_F} = (-1)^{F+F'} \sigma^{\mathrm{o}nLJFM_F}_{nLJ'F'M_F}. \tag{17}$$

In the CSIC model of electron capture a completely diagonal density matrix in the u.a.m. representation transforms to a density matrix in the c.a.m. basis containing four kinds of elements:

(1) Completely diagonal elements,

$$\sigma^{\mathrm{o}nLJFM_F}_{nLJFM_F} \equiv P^{\mathrm{o}}\left(n^2L_jFM_F\right)$$
$$= \sum_{M_L, M_S, M_I} \sigma^{LM_L}_n \left[\langle LSM_LM_S \mid Jm_J\rangle\langle JIM_JM_I \mid FM_F\rangle\right]^2, \tag{18a}$$

which give the initial occupation probability $P^\circ(n^2 L_J F M_F)$ of each state.

(2) Elements diagonal in n, L, J, M_F,

$$\sigma^{\circ nLJFM_F}_{nLJF'M_F} = \sum_{M_L, M_S, M_I} \sigma_n^{LM_L} \langle LSM_L M_S \mid JM_J \rangle^2$$

$$\times \langle JIM_J M \mid FM_F \rangle \langle JIM_J M \mid F'M_F \rangle. \quad (18b)$$

These elements express coherence between hyperfine states within the same fine structure multiplet.

(3) Elements diagonal in n, L, F, M_F,

$$\sigma^{\circ nLJFM_F}_{nLJ'FM_F} = \sum_{M_L, M_S, M_I} \sigma_n^{LM_L} \langle LSM_L M_S \mid JM_J \rangle \langle LSM_L M_S \mid J'M_J \rangle$$

$$\times \langle JIM_J M \mid FM_F \rangle \langle J'IM_J M \mid FM_F \rangle. \quad (18c)$$

These elements express the coherence of excitation between hyperfine structure levels arising from the different fine structure multiplets.

(4) Elements diagonal in n, L, M_F, as expressed by the general relation (14) with $J \neq J'$ and $F \neq F'$. These elements express coherence between hyperfine states in different hyperfine and fine structure multiplets.

For an electronic manifold of given principal quantum number n, there is a total of

$$(2I + 1)(2S + 1) \sum_{L=0}^{n-1} (2L + 1) = 4n^2 \quad (19)$$

different states, and hence $16n^4$ possible density matrix elements. Of this total, $4n^2$ are diagonal elements, and only half of the off-diagonal elements are independent since $\sigma^\circ_{\alpha\beta} = (\sigma^\circ_{\beta\alpha})^*$. Thus, to characterize fully an electronic manifold of hydrogenic states requires $2n^2(4n^2 + 1)$ different elements. Within the framework of the CSIC model, however, the entire manifold can be represented in terms of

$$\sum_{L=0}^{n-1} (L + 1) = \tfrac{1}{2}n(n + 1) \quad (20)$$

distinct capture parameters $\sigma_n^{LM_L}$.

To appreciate this economy of representation, consider the case of the hydrogen $n = 4$ manifold. There are 64 different states and hence a possibility of 2080 distinct density matrix elements. So large a number of independent parameters would be very difficult, if not impossible, to determine from a "reasonable" quantity of data such as furnished by conventional scattering experiments or the electric resonance experiments discussed in this book. In terms of the CSIC model, however, the elements of the $n = 4$ density matrix are fully specified with a set $\{\sigma_n^{LM_L}\}$ of 10 parameters. Normalization of the density matrix by setting $\text{Tr}\{\sigma^\circ\} = 1$ or by defining $\sigma_n^{00} = 1$ reduces this number to 9 relative parameters.

The probability $P^\circ(n^2 L_J F M_F)$ of occupying a state $|nLJFM_F\rangle$ is given by the diagonal density matrix element (18a). Table 8.3 summarizes the initial relative populations of all states with $0 \leq L \leq 3$ in terms of the parameters $\sigma_n^{LM_L}$. Correspondingly, table 8.4 shows hyperfine coherence terms that result from the basis transformation (14). As a check based on the orthonormality properties of the Clebsch-Gordan (or 3-J) symbols, note that if all scattering parameters $\sigma_n^{LM_L}$ are equal, the relative populations of table 8.3 reduce to unity, and the coherences of table 8.4 all vanish.

To obtain the probability of populating a fine structure level without regard to the specific distribution over hyperfine states, sum the elements $P^\circ(n^2 L_J F M_F)$ over all F, M_F quantum numbers. This is facilitated by the orthonormality relations of the 3-J symbols, Eqs. (3.54a, b), and leads to

$$P^\circ(n^2 L_J) = \sum_{F, M_F} P^\circ(n^2 L_J F M_F) = \frac{2(2J + 1)}{(2L + 1)} \sigma_n^L \qquad (21a)$$

where

$$\sigma_n^L \equiv \sum_{M_L} \sigma_n^{LM_L}. \qquad (21b)$$

From Eq. (21a) and the fact that for hydrogen $J = L \pm \frac{1}{2}$ (for $L \neq 0$) and $J = \frac{1}{2}$ (for $L = 0$), it follows that the total probability of occupying a particular orbital level L is

$$P^\circ(n^2 L) = \sum_J P^\circ(n^2 L_J) = 4\sigma_n^L. \qquad (22)$$

Initial orbital populations for $0 \leq L \leq 3$ are summarized in table 8.5.

TABLE 8.3

Initial State Populations

L	J	F	M_F	$P^0(nLJFM_F)$	L	J	F	M_F	$P^0(nLJFM_F)$
0	1/2	1	1	σ_n^{00}	2	3/2	2	0	$\frac{3}{5}\sigma_n^{21} + \frac{2}{5}\sigma_n^{20}$
0	1/2	1	0	σ_n^{00}	2	3/2	1	1	$\frac{3}{5}\sigma_n^{22} + \frac{3}{10}\sigma_n^{21} + \frac{1}{10}\sigma_n^{20}$
0	1/2	0	0	σ_n^{00}	2	3/2	1	0	$\frac{3}{5}\sigma_n^{21} + \frac{2}{5}\sigma_n^{20}$
1	3/2	2	2	σ_n^{11}	3	7/2	4	4	σ_n^{33}
1	3/2	2	1	$\frac{1}{2}\sigma_n^{11} + \frac{1}{2}\sigma_n^{10}$	3	7/2	4	3	$\frac{1}{4}\sigma_n^{33} + \frac{3}{4}\sigma_n^{32}$
1	3/2	2	0	$\frac{1}{3}\sigma_n^{11} + \frac{2}{3}\sigma_n^{10}$	3	7/2	4	2	$\frac{1}{28}\sigma_n^{33} + \frac{3}{7}\sigma_n^{32} + \frac{15}{28}\sigma_n^{31}$
1	3/2	1	1	$\frac{5}{6}\sigma_n^{11} + \frac{1}{6}\sigma_n^{10}$	3	7/2	4	1	$\frac{3}{28}\sigma_n^{32} + \frac{15}{28}\sigma_n^{31} + \frac{5}{14}\sigma_n^{30}$
1	3/2	1	0	$\frac{1}{3}\sigma_n^{11} + \frac{2}{3}\sigma_n^{10}$	3	7/2	4	0	$\frac{3}{7}\sigma_n^{31} + \frac{4}{7}\sigma_n^{30}$
1	1/2	1	1	$\frac{2}{3}\sigma_n^{11} + \frac{1}{3}\sigma_n^{10}$	3	7/2	3	3	$\frac{25}{28}\sigma_n^{33} + \frac{3}{28}\sigma_n^{32}$
1	1/2	1	0	$\frac{2}{3}\sigma_n^{11} + \frac{1}{3}\sigma_n^{10}$	3	7/2	3	2	$\frac{3}{28}\sigma_n^{33} + \frac{5}{7}\sigma_n^{32} + \frac{5}{28}\sigma_n^{31}$
1	1/2	0	0	$\frac{2}{3}\sigma_n^{11} + \frac{1}{3}\sigma_n^{10}$	3	7/2	3	1	$\frac{5}{28}\sigma_n^{32} + \frac{17}{28}\sigma_n^{31} + \frac{3}{14}\sigma_n^{30}$
2	5/2	3	3	σ_n^{22}	3	7/2	3	0	$\frac{3}{7}\sigma_n^{31} + \frac{4}{7}\sigma_n^{30}$
2	5/2	3	2	$\frac{1}{3}\sigma_n^{22} + \frac{2}{3}\sigma_n^{21}$	3	5/2	3	3	$\frac{6}{7}\sigma_n^{33} + \frac{1}{7}\sigma_n^{32}$
2	5/2	3	1	$\frac{1}{15}\sigma_n^{22} + \frac{8}{15}\sigma_n^{21} + \frac{2}{5}\sigma_n^{20}$	3	5/2	3	2	$\frac{1}{7}\sigma_n^{33} + \frac{13}{21}\sigma_n^{32} + \frac{5}{21}\sigma_n^{31}$
2	5/2	3	0	$\frac{2}{5}\sigma_n^{21} + \frac{3}{5}\sigma_n^{20}$	3	5/2	3	1	$\frac{5}{21}\sigma_n^{32} + \frac{10}{21}\sigma_n^{31} + \frac{2}{7}\sigma_n^{30}$
2	5/2	2	2	$\frac{13}{15}\sigma_n^{22} + \frac{2}{15}\sigma_n^{21}$	3	5/2	3	0	$\frac{4}{7}\sigma_n^{31} + \frac{3}{7}\sigma_n^{30}$
2	5/2	2	1	$\frac{2}{15}\sigma_n^{22} + \frac{2}{3}\sigma_n^{21} + \frac{1}{5}\sigma_n^{20}$	3	5/2	2	2	$\frac{5}{7}\sigma_n^{33} + \frac{5}{21}\sigma_n^{32} + \frac{1}{21}\sigma_n^{31}$
2	5/2	2	0	$\frac{2}{5}\sigma_n^{21} + \frac{3}{5}\sigma_n^{20}$	3	5/2	2	1	$\frac{10}{21}\sigma_n^{32} + \frac{8}{21}\sigma_n^{31} + \frac{1}{7}\sigma_n^{30}$
2	3/2	2	2	$\frac{4}{5}\sigma_n^{22} + \frac{1}{5}\sigma_n^{21}$	3	5/2	2	0	$\frac{4}{7}\sigma_n^{31} + \frac{3}{7}\sigma_n^{30}$
2	3/2	2	1	$\frac{1}{5}\sigma_n^{22} + \frac{1}{2}\sigma_n^{21} + \frac{3}{10}\sigma_n^{20}$					

An example of the sensitivity of electric resonance lineshapes to initial state populations may be seen in the experimental lineshapes and computer simulations of figure 8.21 for transitions involving $4F$ and $4D$ states. A spectrometer configuration was chosen so that nearly all P states produced in the target chamber decayed before entering the rf interaction region. The close match between theory and experiment for each target of the figure was obtained with initial orbital populations $\sigma_n^{LM_L}$ comprising $M_L = 0$ substates only, the axis of quantization defined in this case by the rf field and not by the beam. This suggests that,

TABLE 8.4

Initial Hyperfine Coherences

L	J	F	F'	M_F	$\sigma^{\,onLJFM_F}_{\,nLJF'M_F}$
1	3/2	2	1	1	$\frac{\sqrt{3}}{6}\left(\sigma_n^{11} - \sigma_n^{10}\right)$
2	5/2	3	2	2	$\frac{2\sqrt{5}}{15}\left(\sigma_n^{22} - \sigma_n^{21}\right)$
2	5/2	3	2	1	$\frac{\sqrt{2}}{45}\left(\frac{1}{3}\sigma_n^{22} + \frac{2}{3}\sigma_n^{21} - \sigma_n^{20}\right)$
2	3/2	2	1	1	$\frac{\sqrt{3}}{5}\left(\sigma_n^{22} - \frac{1}{2}\sigma_n^{21} - \frac{1}{2}\sigma_n^{20}\right)$
3	7/2	4	3	3	$\frac{2\sqrt{7}}{28}\left(\sigma_n^{33} - \sigma_n^{32}\right)$
3	7/2	4	3	2	$\frac{\sqrt{3}}{140}\left(\frac{1}{5}\sigma_n^{33} + \frac{4}{5}\sigma_n^{33} - \sigma_n^{31}\right)$
3	7/2	4	3	1	$\frac{\sqrt{15}}{56}\left(\frac{1}{2}\sigma_n^{32} + \frac{1}{2}\sigma_n^{31} - \sigma_n^{30}\right)$
3	5/2	3	2	2	$\frac{\sqrt{5}}{7}\left(\sigma_n^{33} - \frac{2}{3}\sigma_n^{32} - \frac{1}{3}\sigma_n^{31}\right)$
3	5/2	3	2	1	$\frac{\sqrt{2}}{105}\left(\sigma_n^{32} - \frac{2}{5}\sigma_n^{31} - \frac{3}{5}\sigma_n^{30}\right)$

for the energy and targets employed, electrons were captured preferentially into states with no component of angular momentum perpendicular to the beam. Application of the rotational transformations discussed in chapter 3 then leads to initial states with equal contributions of $\pm M_L$ components defined with respect to the beam axis, as follows:

$$|nD0\rangle_x = \frac{3}{\sqrt{8}}\left(|nD2\rangle_z - |nD\bar{2}\rangle_z\right) - \frac{1}{2}|nD0\rangle_z,$$

$$|nF0\rangle_x = \sqrt{\frac{5}{16}}\left(|nF3\rangle_z - |nF\bar{3}\rangle_z\right) + \frac{\sqrt{3}}{4}\left(|nF1\rangle_z - |nF\bar{1}\rangle_z\right).$$

TABLE 8.5

Initial Orbital Populations

L	J	$P^o(n^2L_J)$
0	1/2	$4\sigma_n^0$
1	1/2	$\frac{4}{3}\sigma_n^1$
1	3/2	$\frac{8}{3}\sigma_n^1$
2	3/2	$\frac{8}{5}\sigma_n^2$
2	5/2	$\frac{12}{5}\sigma_n^2$
3	5/2	$\frac{12}{7}\sigma_n^3$
3	7/2	$\frac{16}{7}\sigma_n^3$

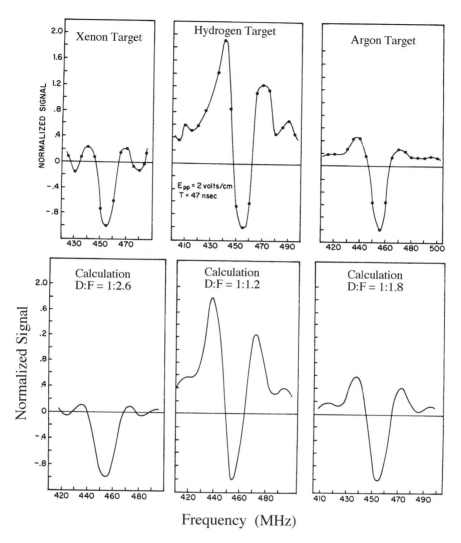

Figure 8.21 Sensitivity of rf lineshapes to initial state populations in the case of hydrogen $4^2F_{5/2}$–$4^2D_{3/2}$ and $4^2D_{5/2}$–$4^2P_{3/2}$ transitions. Top row: experimental spectra recorded with beam energy 18 keV, rf field strength 1 V/cm, rf interaction time 47 ns, and rf propagation angle (relative to beam) π radians. Bottom row: theoretical spectra based on the indicated initial $D{:}F$ population ratios with $M_L = 0$ substates only (defined relative to the rf field axis).

From such linear combinations, scattering parameters defined with respect to the beam axis can be deduced.

In the analysis of initial populations by means of electric resonance theory, computer simulation provides a relatively rapid way to obtain an approximate set of scattering parameters. If the number of sought-for parameters is not too large, they—or certain uncorrelated linear combinations of them to which the spectra are sensitive—can be determined by standard fitting procedures such as the method of least squares.

APPENDIX 8A: THE PARAXIAL RAY EQUATION FOR IONS

The paraxial ray equation relates the radial movement of a charged particle in an electrostatic lens to its axial displacement. For a cylindrically symmetric potential $V(r, z)$ within the lens, the equation of motion of the focused particle takes the form

$$\ddot{r} = -\zeta \frac{\partial V(r, z)}{\partial r} \tag{8A.1}$$

where $\zeta = q/m$ is the charge/mass ratio of the ion, and each dot above a variable indicates differentiation with respect to time; thus, $\ddot{r} \equiv d^2r/dt^2$. (Recall that primes signify differentiation with respect to z.)

Although the demonstation will not be given here, $V(r, z)$, as a solution to Laplace's equation, can be expressed by the integral[20]

$$V(r, z) = \frac{1}{2\pi} \int_0^{2\pi} V(z + ir \sin \alpha) \, d\alpha. \tag{8A.2}$$

Expansion of this integral in a Taylor series about the axis ($r = 0$) and truncation after the second term (for trajectories close to the symmetry axis) lead to the approximate relation

$$V(r, z) \approx V_{0z} - \tfrac{1}{4} r^2 V_{0z}''. \tag{8A.3}$$

The right side of (8A.3) employs notation indicative of an array of discrete points, since potential values are obtained by the relaxation method discussed earlier. Substituting Eq. (8A.3) into (8A.1) and using the chain rule to take derivatives, we obtain

$$\ddot{r} = r'' \dot{z}^2 + r' \ddot{z} = \tfrac{1}{2} \zeta r V_{0z}''. \tag{8A.4}$$

To evaluate $\dot{z}^2$ and $\ddot{z}$ in terms of V_{0z}, derivatives of V_{0z}, and the acceleration potential V_{ac}, we use the conservation of energy relation

$$\tfrac{1}{2}mv(r, z)^2 + qV(r, z) = qV_{ac}, \qquad (8A.5)$$

where $v(r, z) = (\dot{z}^2 + \dot{r}^2)^{1/2} \approx \dot{z}$ is the particle velocity within the lens. From Eq. (8A.5) then follow the relations

$$\dot{z}^2 \approx 2\zeta(V_{ac} - V_{0z}), \qquad (8A.6)$$

$$\ddot{z} \approx -\zeta V_{0z}', \qquad (8A.7)$$

which, when substituted into Eq. (8A.4) yield the paraxial ray equation

$$\ddot{r} + \tfrac{1}{2}r'\left(\frac{V_{0z}'}{V_{0z} - V_{ac}}\right) + \tfrac{1}{4}r\left(\frac{V_{0z}''}{V_{0z} - V_{ac}}\right) = 0. \qquad (8A.8)$$

Equation (8A.8) is strictly valid only for an ion trajectory close to the lens axis. In the absence of an accelerating potential ($V_{ac} = 0$), the equation reduces to the familiar ray equation employed in electron optics.

APPENDIX 8B: EFFECT OF STANDING WAVES
ON A RESONANCE LINESHAPE

The simultaneous interaction of an atom with two rf fields, one much stronger than the other, is easily treated within the resolvent operator formalism of chapter 4 in which the atom and fields constitute a single quantized system. Recall that the state vector of the total system is given by

$$\Psi(t) = \frac{1}{2\pi i} \oint_C dE\, e^{-iEt} G(E)\Psi^0, \qquad (8B.1)$$

where $G(E) = (E - \mathscr{H})^{-1}$ is the resolvent or Green's operator and Ψ^0 is the state vector at time $t = 0$. The contour, C, over which the integration is performed is chosen so as to conform to the principle of causality (as discussed in section 4.7).

In the present case, the total Hamiltonian $\mathcal{H} = \mathcal{H}_0 + \mathcal{H}_1 + \mathcal{H}_2$ is the sum of the unperturbed Hamiltonian of the atom and fields of frequencies ω_1 and ω_2,

$$\mathcal{H}_0 = \sum_\mu (\omega_\mu - \tfrac{1}{2} i\gamma_\mu)|\mu\rangle\langle\mu| + \omega_1 a_1^\dagger a_1 + \omega_2 a_2^\dagger a_2, \qquad (8B.2)$$

and a Hamiltonian characterizing the electric-dipole interaction of the atom with each field,

$$\mathcal{H}_1 = \frac{\vartheta_1}{\sqrt{\bar{n}_1}} p_z(a_1^\dagger + a_1), \qquad (8B.3)$$

$$\mathcal{H}_2 = \frac{\vartheta_2}{\sqrt{\bar{n}_2}} p_z(a_2^\dagger + a_2). \qquad (8B.4)$$

The operator p_z is the z-component of the electron linear momentum, and coupling constants ϑ_1 and ϑ_2 are proportional to the magnitude of the corresponding classical vector potentials. As discussed previously, a classical rf field of mean photon occupation number $\bar{n}$ can be adequately represented by a ket $|n\rangle$ in Fock space for $n \sim \bar{n}$ if phase information is not sought. Utilizing the notation of chapter 4, we designate a state of the total system by the ket $|\mu; q_1, q_2\rangle$, signifying an atom in state $|\mu\rangle$ and a radiation field with q_1 photons of radiofrequency ω_1 and q_2 photons of radiofrequency ω_2.

The above quantal description can be correlated with the classical experimental parameters of section 8.5 as follows. $\mathcal{H}_1$ effects the interaction of the atoms with the strong incident rf field of amplitude $E^{(-)}$ and frequency $\omega_1 = \omega(1 - v/c)$ where ω is the laboratory oscillator frequency and v is the speed of the atoms with respect to the laboratory. $\mathcal{H}_2$ effects the interaction of the atoms with the weak reflected field of amplitude $E^{(+)} = \rho E^{(-)}$ and frequency $\omega_2 = \omega(1 + v/c)$. The process we wish to study is the coupling of an atomic state $|\alpha\rangle$ to an atomic state $|\beta\rangle$ via the stimulated emission or absorption of a single photon of (angular) frequency ω_1 in the presence of photons of frequency ω_1 and ω_2. The method is immediately generalizable to many atomic states, but we concentrate on two since the problem is readily solved and illustrative of the principles involved.

In the analysis to follow we work within the subspace of states projected out by the operator

$$P = |\alpha; n_1, n_2\rangle\langle\alpha; n_1, n_2| + |\beta; n_1 + 1, n_2\rangle\langle\beta; n_1 + 1, n_2|; \quad (8B.5)$$

the reflected field is presumed sufficiently weak that the probability of generating states $|\beta; n_1, n_2 + 1\rangle$ and $|\beta; n_1 + 1, n_2 + 1\rangle$ by real transitions can be neglected. At first thought it may seem that the reflected rf wave has no effect in this approximation since it induces no transitions between the two atomic states. We will see, however, that the presence of this wave is sufficient to produce a small asymmetric shift in the lineshape predicted for a single oscillating field.

The operator $G(E)$, projected onto the space of closely coupled states of interest, takes the form

$$PG(E)P = (E - P\mathcal{H}_0 P - PRP)^{-1}, \quad (8B.6)$$

where

$$\begin{aligned}
\mathcal{R} \equiv PRP &= P(\mathcal{H}_1 + \mathcal{H}_2)P \\
&+ P(\mathcal{H}_1 + \mathcal{H}_2)P[E - P\mathcal{H}_0 P - Q(\mathcal{H}_1 + \mathcal{H}_2)Q]^{-1}P(\mathcal{H}_1 + \mathcal{H}_2)P
\end{aligned}$$
$$(8B.7a)$$

and $Q = 1 - P$ is the complementary projection operator. To utilize Eq. (8B.7a), it is adequate to retain only the first two terms of a Taylor series expansion,

$$\mathcal{R} \approx P(\mathcal{H}_1 + \mathcal{H}_2)P + P(\mathcal{H}_1 + \mathcal{H}_2)\frac{Q}{E - \mathcal{H}_0}(\mathcal{H}_1 + \mathcal{H}_2)P. \quad (8B.7b)$$

The off-diagonal elements of $\mathcal{R}$ within the restricted subspace of Eq. (8B.5) are independent of the interaction $\mathcal{H}_2$. The diagonal elements, however, which are responsible for the level shifts, do depend on $\mathcal{H}_2$, and lead to a lineshape displacement

$$\delta\Omega = \mathrm{Re}\{\mathcal{R}_{\alpha; n_1, n_2} - \mathcal{R}_{\beta; n_1+1, n_2}\} \quad (8B.8)$$

that differs from the shift arising from the interaction with a single electric field. The complete derivation of the wavefunction and resonance lineshapes can be carried out as described in earlier chapters. Here we concern ourselves solely with the resonance shift which, to the level of approximation we are considering, is the only new feature. Evaluation of the two terms in Eq. (8b.8) yields

$$\mathcal{R}_{\alpha; n_1, n_2} = \frac{|V_1|^2}{E - E_{\beta; n_1-1, n_2}} + \frac{|V_2|^2}{E - E_{\beta; n_1, n_2+1}}$$
$$+ \frac{|V_2|^2}{E - E_{\beta; n_1, n_2-1}}, \tag{8B.9a}$$

and

$$\mathcal{R}_{\beta; n_1+1, n_2} = \frac{|V_1|^2}{E - E_{\alpha; n_1+2, n_2}} + \frac{|V_2|^2}{E - E_{\alpha; n_1+1, n_2+1}}$$
$$+ \frac{|V_2|^2}{E - E_{\alpha; n_1+1, n_2-1}}, \tag{8B.9b}$$

where $V_k = \langle \alpha | \vartheta_k p_z | \beta \rangle$ $(k = 1, 2)$, and $E_{\mu; q_1, q_2} = (\omega_\mu - \frac{1}{2} i \gamma_\mu) + q_1 \omega_1 + q_2 \omega_2$. When the field $E^{(-)}(t)$ is tuned to resonance, we have $\omega_1 \approx \omega_0 \equiv \omega_\alpha - \omega_\beta$, or $\omega \approx \omega_0 (1 - v/c)^{-1} \approx \omega_0 (1 + v/c)$, and E can be approximated by $E \approx E_{\alpha; n_1, n_2}$. Substituting these results into relations (8B.9a, b) and retaining the most significant contributions lead to the shift

$$\delta \Omega = \frac{|V_1|^2}{\omega_0} + \frac{2|V_2|^2(\omega_0 - \omega_2)}{(\omega_0 - \omega_2)^2 + \frac{1}{4}(\gamma_\alpha - \gamma_\beta)^2} + \frac{2|V_2|^2}{\omega_0 + \omega_2} \tag{8B.10a}$$

or equivalently

$$\delta \Omega = \frac{|V_1|^2}{\omega_0} \left(1 + \frac{|\rho|^2 \frac{v}{c}}{\frac{v^2}{c^2} + \left(\frac{\gamma_\alpha - \gamma_\beta}{4\omega_0} \right)^2} + \frac{|\rho|^2}{\left(1 + \frac{v}{2c} \right)} \right). \tag{8B.10b}$$

The first term of Eqs. (8B.10a, b) is the ac Stark shift derived in earlier chapters. The second term depends on the reflection coefficient and exhibits a first-order Doppler effect. The third term also depends on the reflection coefficient, but is largely insensitive to the Doppler effect, reducing to $|V_1|^2 |\rho|^2 / \omega_0$ for $v/c \ll 1$.

When the rf field propagates perpendicular to the atomic beam, the frequency shift, with neglect of the second-order Doppler effect, consists of two terms (in effect, the first and third of the preceding equation)

$$\delta\Omega = \frac{|V_1|^2}{\omega_0}(1 + |\rho|^2).$$ (8B.11)

If the reflected wave is intense enough to induce transitions between atomic states $|\alpha\rangle$ and $|\beta\rangle$, the projection operator must include the additional term $|\beta; n_1, n_2 + 1\rangle\langle\beta; n_1, n_2 + 1|$. The true resonance lineshape is then obtained from Eq. (8B.1) with substitution of the full 3×3 Green's matrix into the integrand.

APPENDIX 8C: PHENOMENOLOGICAL MODEL OF THE RF CHAMBER

Figure 8.22 shows a schematic diagram of the rf chamber in which each connector is represented as a two-port device inserted into the transmission line. We will analyze this system in terms of conventional circuit theory variables, instead of electromagnetic fields. Thus, the incident wave, represented by a potential V_1^+, ideally propagates through the chamber unperturbed (except for interaction with the atomic beam). When impedance mismatch occurs, however, there are back-propagating waves, and the potential seen by an atom, $V_2^+ + V_2^-$, can differ from the potential at the output terminal, $V_3^+ + V_3^-$. The objective of the model is to relate the power in the interaction region to the measured power output by means of experimentally determinable scattering matrix elements. This permits the experimenter to correct for any power variation with frequency. It is assumed throughout that the power varies smoothly with frequency over the frequency range in which a rf chamber is to be used.

The waves reflected at the first and second two-port devices are related to the waves incident at these devices by the scattering matrices S and R respectively through the equations

$$\begin{pmatrix} V_1^- \\ V_2^- \end{pmatrix} = \begin{pmatrix} S_{11} & S_{12} \\ S_{21} & S_{22} \end{pmatrix} \begin{pmatrix} V_1^+ \\ V_2^+ \end{pmatrix},$$ (8C.1a)

$$\begin{pmatrix} V_2^+ \\ V_3^- \end{pmatrix} = \begin{pmatrix} R_{11} & R_{12} \\ R_{21} & R_{22} \end{pmatrix} \begin{pmatrix} V_2^- \\ V_3^+ \end{pmatrix},$$ (8C.1b)

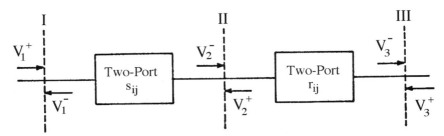

Figure 8.22 Model of the rf chamber with each connector represented as a two-port device inserted into the transmission line. I, II, and III mark reference planes before, within, and after the rf chamber at which the indicated potentials are measured.

where the potentials are defined at the reference planes indicated in the figure. The elements of the scattering matrices are complex numbers characterizing each device in terms of its effect on the incident rf wave. In solving Eqs. (8C.1a, b), we are justified in making the following simplifying assumptions based on the geometry or material properties of the rf chamber:

(1) *Electrical symmetry.* As a consequence of the bilateral symmetry of the chamber design, the microwave transmission is unaltered by an end-to-end reversal of the chamber position in the circuit. Thus

$$S_{ij} = R_{ij} \qquad (i, j = 1/2). \qquad (8C.2)$$

In addition, since the rf connectors are small in comparison to a wavelength, a reversal of the connector direction also leaves the transmission line unaltered. Hence

$$S_{11} = S_{22},$$
$$R_{11} = R_{22}. \qquad (8C.3)$$

(2) *Reciprocity.* A device is reciprocal if it contains only isotropic media and its parameters do not change with the magnitudes of the fields. (Among such devices are connectors, transmission lines, attenuators, and

filters; not included are amplifiers, mixers, or ferrite isolators.) This implies that

$$Z_2 S_{12} = Z_1 S_{21},$$
$$Z_1 R_{12} = Z_2 R_{21},$$

(8C.4)

where Z_1 is the characteristic impedance in regions I and III, and Z_2 is the characteristic impedance in region II. The characteristic impedances are assumed equal in this model.

(3) *Losslessness.* Under the condition (met by a well-designed rf chamber) that no energy is internally dissipated by resistive losses, the scattering matrix elements at each port satisfy the additional relations

$$|S_{12}| = |S_{21}| = \sqrt{1 - |S_{11}|^2},$$
$$|R_{12}| = |R_{21}| = \sqrt{1 - |R_{11}|^2}.$$

(8C.5)

(4) *Matched Load.* If the terminal impedance is equal to the characteristic impedance of the transmission line, then there is no reflection from the load (thermistor), and one can set $V_3^+ = 0$.

Taking the foregoing approximations into account, we define a reflection coefficient $\rho = |\rho| e^{i\phi} = S_{11}$ and transmission coefficient $\tau = S_{12}$ subject to the lossless condition $|\rho|^2 + |\tau|^2 = 1$. The matrix equations then lead to the ratios

$$\frac{V_1^-}{V_1^+} = \frac{\rho(1 + \tau^2 - \rho^2)}{1 - \rho^2},$$

(8C.6a)

$$\frac{V_2^-}{V_1^+} = \frac{\tau}{1 - \rho^2},$$

(8C.6b)

$$\frac{V_2^+}{V_1^+} = \frac{\rho\tau}{1 - \rho^2},$$

(8C.6c)

$$\frac{V_3^-}{V_1^+} = \frac{\tau^2}{1 - \rho^2}.$$

(8C.6d)

The total potential V_a experienced by the atom is, from figure 8.22, the superposition of the incident and reflected waves within the chamber:

$$V_a = V_2^+ + V_2^- = V_3^- \left(\frac{1 + \rho}{\tau} \right).$$

(8C.7)

It then follows that the power in the interaction region,

$$P_a = \text{constant} \times |V_a|^2, \tag{8C.8}$$

and the measured power at the load,

$$P_L = \text{constant} \times |V_3^-|^2, \tag{8C.9}$$

are related by the expression

$$P_a = P_L \left[\frac{1 + |\rho|^2 + 2|\rho| \cos \phi}{1 - |\rho|^2} \right]. \tag{8C.10}$$

To determine the phase in Eq. (8C.10) is not usually a simple matter, but the magnitude of the reflection coefficient, in the case where ρ is small and $\tau \sim 1$, can be written in terms of the measurable quantities $|V_1^+|$ (the power to the load in the absence of the rf chamber), $|V_3^-|$ (the power to the load in the presence of the rf chamber), and the voltage standing wave ratio (SWR) in region I, defined by

$$\sigma \equiv \frac{|V_1|_{\text{max}}}{|V_1|_{\text{min}}} = \frac{1 + (|V_1^-|/|V_1^+|)}{1 - (|V_1^-|/|V_1^+|)}. \tag{8C.11}$$

We then find that

$$|\rho| = -\frac{|V_1^-|/|V_1^+|}{1 + (|V_3^-|/|V_1^+|)} = \frac{(\sigma - 1)/(\sigma + 1)}{1 + (|V_3^-|/|V_1^+|)}. \tag{8C.12}$$

For rf chambers in which the power correction is very small, Eq. (8C.6a) shows that $|V_1^-|/|V_1^+|$ is virtually identical to $2|\rho|$.

NOTES

1. J. H. Moore, C. C. Davis, and M. A. Coplan, *Building Scientific Apparatus* (Addison-Wesley, Reading, MA, 1983), chap. 3.

2. M. P. Silverman, "Zeeman and Stark Effects," in *Encyclopedia of Applied Physics*, vol. 23, ed. by G. Trigg (Wiley–VCH Verlag, New York, 1998), pp. 563–585.

3. J. D. Cockcroft and E. T. S. Walton, "Experiments with High Velocity Positive Ions," *Proc. Roy. Soc. (London)* **A129** (1930):477.

4. R. W. Wood, *Proc. Roy. Soc. (London)* **A97** (1920):455.

5. J. J. and G. P. Thomson, *Conduction of Electricity through Gases*, Vol. II (Cambridge University Press, London, 1928). The father-son pair represent an interesting historical duality, the former having revealed the existence of the electron as a particle, and the latter having shown the electron to behave as a wave.

6. C. D. Moak, H. Reese, Jr., and W. M. Good, "Design of a Radio-Frequency Ion Source for Particle Accelerators," *Nucleonics* **9** (September 1951):18.

7. D. Blanc and A. Degeilh, *J. Physique et Radium* **22** (1961):230.

8. P. W. Chudleigh, *Rev. Sci. Instr.* **39** (1968):356.

9. The deflection of an ion passing from a region of electrostatic potential V_1 to a region of potential V_2 is described by a relation identical in form to Snel's law in light optics: $\sin \vartheta_1 / \sin \vartheta_2 = \sqrt{V_2}/\sqrt{V_1}$, in which ϑ_1 and ϑ_2 are the angles of incidence and refraction measured with respect to the normal line. Thus, one can take the refractive index of a region to be proportional to the square root of the electrostatic potential. Note that ions move faster the higher the refractive index (the voltage), whereas the phase velocity of a light wave is slower in a medium of higher refractive index. When magnetic fields are present, the ion refractive index will also depend on the vector potential (which can lead to counterintuitive quantum interference effects such as the Aharonov-Bohm effect). See. for example, M. P. Silverman, *And Yet It Moves, Strange Systems and Subtle Questions in Physics* (Cambridge University Press, New York, 1993) and *More than One Mystery: Explorations in Quantum Interference* (Springer, New York, 1995).

10. Let v_1 be the proton speed at the extraction tip and v_2 the speed at the (grounded) tapered lens element. Conservation of energy applied at these two points leads to the relations $qV_{ex} = \frac{1}{2}mv_1^2$ and $q(V_{ac} - V_{ex}) + \frac{1}{2}mv_1^2 = \frac{1}{2}mv_2^2$, from which follows the ratio $v_2^2/v_1^2 = V_{ac}/V_{ex}$.

11. For a good description of the relaxation method see (a) S. Ramo, J. R. Whinnery, and T. van Duzer, *Fields and Waves in Communication Electronics* (Wiley, New York, 1965), pp. 164–168, (b) E. M. Purcell, *Electricity and Magnetism* (McGraw-Hill, New York, 1965), pp. 102–104, 416–418.

12. G. N. Plass, *J. Appl. Phys.* **13** (1942):49.

13. S. Bashkin, W. S. Bickel, D. Fink, and R. K. Wangsness, "Interference of Fine-Structure Levels in Hydrogen," *Phys. Rev. Lett.* **15** (1965):284. I discuss the phenomenon of quantum beats comprehensively in the two books of note 9.

14. These conditions were estimated from the data of S. K. Allison, *Rev. Mod. Phys.* **30** (1958):1137.

15. H. A. Bethe and E. E. Salpeter, *Quantum Mechanics of One- and Two-Electron Atoms* (Springer, Heidelberg, 1957), p. 266.

16. N. F. Ramsey, *Molecular Beams* (Oxford University Press, NY, 1956), pp. 139, 155–156; S. Millman, *Phys. Rev.* **55** (1939):628.

17. C. W. Fabjan, *Resonance Narrowed Lamb Shift Measurement In Hydrogen*, $n = 3$ (Ph.D. Thesis, Harvard University, 1971).

18. M. P. Silverman, "Zeeman and Stark Effects," in *Encyclopedia of Applied Physics* vol. 23 (Wiley–VCH Verlag, New York, 1998), pp. 563–585.

19. M. P. Silverman, *Optical Electric Resonance Investigation of a Fast Hydrogen Beam* (Ph.D. Thesis, Harvard University, 1973); M. P. Silverman and F. M. Pipkin, "Optical Electric Resonance Investigation of a Fast Hydrogen Beam, Part III: Experimental Procedure and Analysis of H($n = 4$) Quantum States," *J. Phys. B: Atom. Molec. Phys.* **7** (1974):747.

20. V. E. Coslett, *Introduction to Electron Optics* (Oxford University Press, London, 1946), section 12.